BEFORE HUMANS

THE DRAMA OF WORLD-MAKING

MICHAEL VINCENT

First Edition, 2026

ISBN: 978-1-971762-91-3

CONTENTS

INTRODUCTION

This book tells the story of how the world was made.

Not the human world—that story comes later. This is the story of what came before: the formation of the cosmos itself, the birth of stars and planets, the preparation of Earth for life, the emergence and evolution of life across billions of years, and the long journey from single-celled organisms to creatures capable of reading these words.

It is a story told many times, in many ways. Scientists tell it through data and equations, tracing the evidence written in starlight and stone. Mythologies tell it through gods and heroes, encoding ancient wisdom about order and chaos, creation and destruction. This book tells it through a different lens—a cosmological text that claims to reveal what actually happened, not through scientific investigation or cultural transmission, but through direct knowledge from beings who participated in the process.

The source text I have drawn upon is unusual. It presents itself as an account of cosmic and planetary history authored by celestial beings —administrators, observers, and other personalities from a highly organized universe. The account it offers is detailed, internally consistent, and surprisingly compatible with much of modern science.

The story it tells is arresting. It describes a cosmos with a center—a place from which gravity radiates and around which all galaxies orbit. It describes the formation of our local region of space through the evolution of a vast nebula, from which our sun was born as one of nearly a million stars. It describes Earth's formation through meteoric accretion, the establishment of a global ocean, and the drift of continents—a claim that was scientifically controversial when the text was composed but is now accepted fact.

Most distinctively, this cosmological framework claims that life did not arise spontaneously on Earth but was deliberately introduced by beings who prepared the planet, implanted the first living cells, and guided evolution toward its ultimate goal: the emergence of human beings capable of knowing their Creator.

This is not a claim science can evaluate. It is a philosophical and theological position. But it is a position that gives the story of life a different shape—not random variation preserved by natural selection, but purposeful development toward a destination.

I have written this book for general readers interested in the great questions of origins.

The original source material is extensive, dense, and written in an early twentieth-century style that can feel inaccessible to contemporary readers. Its cosmological sections are scattered across hundreds of pages, mixed with theological discussions and administrative details that, whatever their importance in the larger work, are not directly relevant to the story of Earth's formation and the evolution of life herein presented.

My goal has been to extract that story and present it in a form that contemporary readers can engage with: clear prose, vivid imagery, and an honest acknowledgment of where the source material aligns with modern science and where it diverges.

I have not hidden the divergences. Where the source text's timeline conflicts with current evidence, I say so. Where its mechanisms differ from established science, I explain both accounts. Where its claims cannot be evaluated empirically, I present them as claims and let readers decide what weight to give them.

The text itself anticipates this. In a remarkable passage, its authors warn future readers:

> *Because your world is generally ignorant of origins, even of physical origins, it has appeared to be wise from time to time to provide instruction in cosmology. And always has this made trouble for the future. The laws of revelation hamper us greatly by their proscription of the impartation of unearned or premature knowledge. Any cosmology presented as a part of revealed religion is destined to be outgrown in a very short time. Accordingly, future students of such a revelation are tempted to discard any element of genuine religious truth it may contain because they discover errors on the face of the associated cosmologies therein presented.*
>
> *We who participate in the revelation of truth are very rigorously limited by the instructions of our superiors. We are not at liberty to anticipate the scientific discoveries of a thousand years. We full well know that, while the historic facts and religious truths of this series of revelatory presentations will stand on the records of the ages to come, within a few short years many of our statements regarding the physical sciences will stand in need of revision in consequence of additional scientific developments and new discoveries. These new developments we even now foresee, but we are forbidden to include such humanly undiscovered facts in the revelatory records. Let it be made clear that revelations are not necessarily inspired. The cosmology of these revelations is not inspired.*[1]

Yet the authors did not consider this a reason for silence. Even an imperfect cosmology, they argued, has value: it reduces confusion, coordinates scattered facts, restores lost knowledge, and illuminates the spiritual teachings it accompanies.

They expected errors to be found. They asked only that readers not discard everything because of them.

But I have also not dismissed the divergences as proof of error. Science has been wrong before. The text's assertion of continental drift was rejected by mainstream geology when the text was composed and is now universally accepted. Its description of a young sun's variable output matches current models of stellar evolution. Its measurements of Earth's density are precisely accurate. Whatever its ultimate origin, the text contains genuine knowledge alongside its more contested claims.

On Methodology

Throughout this book, I refer to my source simply as "the text." This is deliberate. I want readers to engage with the narrative on its own terms, evaluating its claims by their coherence and their correspondence with known facts, rather than through the lens of other associations. An appendix provides more information about the source for readers who wish to explore it further.

The endnotes to each chapter provide sources for quotations and indicate which sections of the source material inform each chapter. Readers who wish to explore further will find a map to the relevant portions. Scientific and historical citations are also included.

What does this book offer that other accounts of cosmic and planetary history do not?

First, a narrative. Popular science books about Earth's history tend to be episodic—collections of interesting facts rather than unified stories. The account I draw upon is genuinely narrative. It has a beginning, a middle, and an end. It traces a continuous thread from the formation of space itself to the emergence of human consciousness. Reading it feels less like studying a textbook and more like following a story.

Second, a sense of purpose. Scientific accounts of cosmic history are, by design, mechanistic. They describe what happened without asking why. The framework presented here is frankly teleological—it asserts that the cosmos is going somewhere, that life is meant to produce something, that the emergence of human consciousness was not an accident but a goal. It gives the story a different emotional texture than accounts that treat existence as meaningless.

Third, an integration of science and spirituality that neither dismisses the former nor reduces the latter. This account does not ask readers to choose between evolution and creation. It affirms evolution as the mechanism by which life developed while asserting that the process was guided by intelligence. It takes scientific knowledge seriously—often demonstrating surprising accuracy on details that later research confirmed—while embedding that knowledge in a framework of cosmic purpose. The story it tells is worth knowing. It is, at its heart, a story about how life came to be.

This book covers the period from the formation of the cosmos to the emergence of the first human beings approximately one million years ago. It does not cover human history, human civilization, or the spiritual and philosophical teachings that inform the source material's larger vision. Those are stories for other books.

What this book offers is a foundation: an account of how the stage was set, how the conditions for life were established, how life emerged and developed, and how the long process of evolution finally produced creatures capable of asking where they came from.

The drama of world-making took billions of years. The drama of human history is still unfolding. My hope is that understanding the first will illuminate the second—that knowing where we came from will help us understand who we are and where we might be going.

The story begins at the center of everything.

THE ARCHITECTURE OF INFINITY

ravity is the omnipotent strand on which are strung the gleaming stars, blazing suns, and whirling spheres which constitute the universal physical adornment of the eternal God.

Begin at the center.

Not the center of our galaxy, though that too is a place of immense power, a supermassive black hole around which four hundred billion stars slowly wheel. Not the center of our local group of galaxies, nor of the supercluster that contains it. Begin instead at the center of everything. The absolute nucleus of reality itself.

Somewhere in the infinite expanse of existence, there is a literal, physical center to all that is. A place so fundamental that space itself does not touch it. A gravitational anchor so powerful that every star, every galaxy, every wisp of cosmic dust in the farthest reaches of creation feels its pull and responds to its presence.

This is the audacious proposition at the heart of the cosmology we will explore: the cosmos has an architecture. It has a nucleus. It has a geography. Far from being a random scattering of matter across an infinite void, the universe is organized around a central point, an eternal Isle from which all energy flows and toward which all gravity pulls.

The prevailing cosmological model of our age denies this. We are told the universe has no center, that a primordial explosion called the Big Bang was not an explosion from a single point but an expansion of space itself, occurring everywhere at once. We are told the cosmos is approximately 13.8 billion years old, that it emerged from a hot, dense state and has been expanding ever since, and that this expansion is now accelerating toward an eventual heat death in which all stars burn out and all matter disperses into cold, dark emptiness.

But the Big Bang model, for all its dominance in textbooks and popular science, is not without serious problems.

It requires us to accept that 95 percent of the universe consists of substances never directly observed: dark matter and dark energy, inferred to explain why galaxy clusters hold together despite their velocities and why the expansion of space appears to be accelerating. When a theory requires 95 percent of reality to be invisible and undetectable, we might reasonably ask whether the theory itself is flawed.

It requires us to accept "inflation," a hypothetical period of exponential expansion in the first fraction of a second after the Big Bang, during which space itself stretched so rapidly that distant regions were carried apart faster than light could have traveled between them. Inflation was proposed to solve three problems: why distant regions of the universe have the same temperature despite never having been in causal contact (the horizon problem), why space appears geometrically flat (the flatness problem), and why the magnetic monopoles predicted by particle physics are nowhere to be found. Inflation is not observed; it is inferred, a patch applied to fix problems the original theory could not solve.

It requires us to accept the "cosmological constant problem," sometimes called the worst prediction in the history of physics: quantum field theory predicts that empty space should contain vastly more energy than we observe—traditional estimates suggest a discrepancy of 10^{120}, though more rigorous calculations place it at 50 to 60 orders of magnitude. Rather than take this as evidence that something is fundamentally wrong, the discrepancy is set aside as a mystery to be solved later.

And now the James Webb Space Telescope is returning images of galaxies in the supposedly early universe that appear more massive and more numerous than standard models predicted. While some initial reports of "impossible" galaxies have since been explained—supermassive black holes were making certain galaxies appear brighter and more massive than they actually were—the data still shows roughly twice as many massive early galaxies as expected, challenging our understanding of how quickly stars formed in the young cosmos.

The Big Bang is not settled science. It is a working hypothesis, increasingly refined by observations it did not predict. The cosmology presented in this book offers a different picture:

Paradise is the geographic center of infinity.

Imagine, for a moment, that you could travel inward. Past the boundaries of our Milky Way galaxy. Past the billions of other galaxies. Past realms of space where even light has not yet had time to travel since the first moments of cosmic existence. Imagine journeying not outward into the expanding frontier, but inward toward the heart of everything. What would you find there?

You would find an Isle. Not of rock and water, but of pure reality. A place that exists outside of time because time itself begins at its borders. A place that is not spherical, like planets and stars, but ellipsoid: one-sixth longer in the north-south diameter than in the east-west diameter. The central Isle is essentially flat, and the distance from the upper surface to the nether surface is one tenth that of the east-west diameter.

These precise dimensions are significant. They establish absolute direction in a cosmos that otherwise has no up or down, no north or south. In a universe where everything is relative, Paradise provides the fixed reference point. Its asymmetry creates orientation. Its shape defines the geometry of all creation.

From its underside, energy pours forth in waves that have never lapsed, never failed, since the first moment of reality. From its upper regions, the infinite presence of the divine source radiates outward through all the realms of created existence. The center does not move. It cannot move. It is the one fixed point around which everything else turns.

> Absolutely nothing is stationary in all the master universe except the very center of Havona, the eternal Isle of Paradise, the center of gravity.

Every galaxy in our observable universe, every particle of matter, every wave of energy traces an impossibly vast circuit that ultimately references this central point.

But what holds it all together? What prevents the galaxies from flying apart, the stars from scattering into the void, the very atoms in your body from dispersing into random particles?

Gravity.

We know gravity well, or think we do. Newton described it as a force of attraction between masses. Einstein reimagined it as the curvature of spacetime itself, matter telling space how to bend, and bent space telling matter how to move. We experience it every moment of our lives, so constantly that we forget it is there until we stumble or drop something.

But gravity is more than a property that matter happens to possess. It is the fundamental organizing principle of the cosmos, a literal connection between every physical thing and the cosmic center from which all things ultimately derive.

> *The inescapable pull of gravity effectively grips all the worlds of all the universes of all space. Gravity is the all-powerful grasp of the physical presence of Paradise.*

The image is of a cosmos bound together, every particle linked by invisible lines of force to a central source. The stars do not merely happen to orbit. They are held. The galaxies do not merely happen to cluster. They are drawn. And the drawing force traces back, ultimately, to a single point of absolute anchoring.

> *Paradise is the absolute source and the eternal focal point of all energy-matter in the universe of universes. Everything is drawn inward towards Paradise.*

This view resolves problems that plague the standard model. If gravity emanates from a cosmic center, the large-scale structure of the universe makes sense. The mysterious "dark matter" invoked to explain galactic dynamics may simply be the pull of this central gravity, misinterpreted by a model that denies its existence.

And consider this: the so-called "axis of evil," a pattern in the cosmic microwave background radiation that appears aligned with our own solar system's ecliptic plane, should be impossible in a universe with no center and no preferred direction. While scientists debate whether this alignment represents a genuine cosmological anomaly or a statistical artifact, its persistence in data from multiple satellites remains unexplained. In a centered cosmos, such alignments would not be anomalies at all. They would be expected features.

The cosmos breathes. Space itself is not static but dynamic, pulsing in cycles of expansion and contraction that span billions of years.

> *For a billion years of Earth time the space reservoirs contract while the master universe and the force activities of all horizontal space expand. It thus requires a little over two billion years to complete the entire expansion-contraction cycle.*

In 1927, the Belgian physicist and Catholic priest Georges Lemaître published a paper describing a relationship between the distances of galaxies and their velocities—the farther away, the faster they receded. Two years later, Edwin Hubble, working with the 100-inch telescope at Mount Wilson above Los Angeles, published more extensive observations confirming the same pattern: the universe was expanding. In 2018, the International Astronomical Union voted to rename "Hubble's Law" the "Hubble-Lemaître Law" in recognition of both contributions.

But the Big Bang model interprets this expansion as the aftermath of a singular creation event, a one-time explosion still playing out. The cosmology presented here interprets it differently: we are witnessing one half of a cosmic breath, the exhalation phase of a cycle that will eventually reverse. The outward thrust will slow, stop, and give way to contraction.

> *All forms of force and all phases of energy seem to be encircuited; they circulate throughout the universes and return by definite routes. The outer zone pulsates in agelong cycles of gigantic proportions. For a little more than one billion Earth years the space-force of this center is outgoing; then for a similar length of time it will be incoming. And the space-force manifestations of this center are universal; they extend throughout all pervadable space.*

This cyclical model has advantages the Big Bang lacks. It does not require a moment of creation from nothing, a philosophical impossibility that physicists typically wave away with mathematics. It does not require inflation or dark energy or any of the patches invented to rescue a model that keeps encountering observations it cannot explain. It describes a cosmos in dynamic equilibrium, eternally pulsing, eternally alive.

The universe is not winding down toward heat death. It is breathing.

Space itself has structure.

Space does not touch Paradise; only the quiescent midspace zones come in contact with the central Isle.

This is a striking image: space, the empty void we think of as infinite nothingness, actually ends somewhere. It has boundaries. At those boundaries, where space gives way to something else entirely, lies the center of everything. The geometry is specific:

The vertical cross section of total space would slightly resemble a Maltese cross, with the horizontal arms representing the space of the universes and the vertical arms representing reservoir space. Between them lie quiet zones, vast elliptical regions of relative motionlessness that separate the moving space levels. You may visualize the first outer space level, where untold universes are now in process of formation, as a vast procession of galaxies swinging around Paradise, bounded above and below by the midspace zones of quiescence.

The galaxies do not move randomly. They process in orderly circuits, their alternate motions providing gravitational stabilization to the whole.

The cosmos is engineered. Its structure is not accidental but designed to maintain stability across inconceivable distances and timescales.

Where, in all of this, is Earth?

We are newcomers. Our sun is a middle-aged star in a young region of a young galaxy in a not fully organized corner of the cosmos. We exist on the outer edge of things, far from the great central regions where reality is oldest and most established.

And yet we are not lost.

Our path through the cosmos is known. Beings of sufficient perspective can chart our trajectory the way we chart the orbits of our own planets. The apparent chaos of our local neighborhood, dust clouds and asteroid belts and the random scatter of nearby stars, is actually part of an orderly procession, a grand circuit that will eventually bring us back through the same regions of space we are passing through now.

We are not adrift. We are not random. We are on a journey with a trajectory that is understood, if not by us, then by those with wider vision.

The numbers are staggering. The distances are incomprehensible. And yet there is something almost comforting in the picture: we have an address. We have a position. We have a trajectory. In a cosmos with a center, we know roughly where we stand in relation to it.

This chapter has sketched an architecture.

A cosmic center from which all energy flows. Gravity as the strand connecting every particle to that center. Space that breathes in cycles of expansion and contraction. Quiet zones separating vast processions of wheeling galaxies. A charted path for our small world through the immensity of it all.

This is not the cosmos of the textbooks, with its centerless expansion from a primordial singularity, its dark matter and dark energy, its heat death billions of years hence. It is an older vision, and perhaps a truer one: a cosmos with structure, with direction, with purpose. A universe that is going somewhere rather than merely dissipating into cold and darkness.

The universe we will explore in the chapters ahead is not a random accident. It is organized. Structure exists at every scale. Patterns repeat across billions of light-years. The same physics that governs the falling of an apple governs the collapse of a gas cloud into a star a hundred million light-years away.

We are about to trace that structure from its grandest expression down to its most intimate. From the cosmic center to the subatomic particle. From the birth of nebulae to the first stirring of life in warm Cambrian seas. From the age of giants, dinosaurs lumbering across prehistoric continents, to the emergence of small, clever creatures who would one day look up at the stars and wonder where they came from.

It is a story that spans billions of years and culminates in this moment: you, reading these words, contemplating your own origins in the heart of the cosmos.

The architecture has been sketched. Now let us descend into the details.[1]

2

THE NATURE OF MATTER

he foundation of the universe is material in the sense that energy is the basis of all existence.

In the previous chapter, we stood at the center of everything and gazed outward at the vast architecture of space. Now we must reverse our direction. In lieu of contemplating the immense, we must examine the infinitesimal. Instead of galaxies wheeling through the void, we must consider the particles that compose them.

For physics revealed a strange truth in the twentieth century: the solid world is not solid. The chair you sit in, the body you inhabit, the ground beneath your feet—all of it is mostly empty space in terms of where the mass resides. If you could remove all the emptiness from every atom in your body and compress what remained to nuclear density, you would fit in a sphere smaller than a grain of sand. The rest is void, held in shape by forces rather than by substance.

And stranger still: at the smallest scales, the distinction between matter and energy dissolves entirely. What we call a particle is also a wave. What we call mass is also frozen energy. Einstein's famous equation, E=mc², tells us that matter and energy are convertible currencies, different denominations of the same cosmic wealth.

Light, heat, electricity, magnetism, chemism, energy, and matter are, in origin, nature, and destiny, one and the same thing, together with other material realities as yet undiscovered on Earth.

The entire material universe is crystallized energy. Every atom, every electron, every particle down to the smallest constituent of reality is ultimately made of the same primordial force-energy that streams eternally from the cosmic center.

This vast stream of energy proceeding from the Paradise Presences has never lapsed, never failed; there has never been a break in the infinite upholding.

Consider what this means. From the first moment of reality until now, through timescales that dwarf human comprehension, energy has flowed continuously from the cosmic center outward into all the realms of space. It flows still. It will flow forever. The universe is not winding down from some primordial explosion, gradually dissipating into entropy and heat death. It is continuously sustained, continuously fed, continuously upheld by an inexhaustible source.

All force is circuited in Paradise, comes from the Paradise Presences and returns thereto.

Energy does not merely flow outward and disperse. It circulates. It completes circuits. It returns to its source even as new energy pours forth. The cosmos is not a one-way flow toward dissolution but an eternal circulation, a breathing in and out on scales of time and space that stagger imagination.

Force-energy is imperishable, indestructible; these manifestations of the Infinite may be subject to unlimited transmutation, endless trans-formation, and eternal metamorphosis; but in no sense or degree, not

Energy cannot be created or destroyed. This is the first law of thermodynamics, confirmed by every experiment physics has ever devised. But here that law is given a larger context. Energy cannot be destroyed because it is the fundamental expression of reality itself, the material manifestation of an eternal source. It transforms, it transmutes, it changes form endlessly, but it never ceases to exist.

Now let us descend through the layers of matter, from the familiar to the strange.

You know atoms. Even if you have forgotten your high school chemistry, you carry an intuitive sense of them: tiny structures, with a dense nucleus of protons and neutrons surrounded by electrons whose positions are described by probability distributions—fuzzy clouds of potential presence rather than neat planetary orbits. Roughly a hundred different types of atoms, the elements of the periodic table, combine in various configurations to produce every substance you have ever touched, tasted, or breathed.

You may know that atoms themselves are not fundamental. Protons and neutrons are made of still smaller particles called quarks, bound together by the strong nuclear force. The modern Standard Model of particle physics describes a zoo of fundamental particles: six types of quarks, six types of leptons including the electron, and a collection of force-carrying bosons that mediate the four fundamental interactions.

This is the current scientific picture, affirmed by decades of experiments at particle accelerators around the world. The electron, in this framework, is fundamental. It has no internal structure, no component parts. It simply is.

But there is something deeper still.

The ultimatons, unknown on Earth, slow down through many phases of physical activity before they attain the revolutionary-energy prerequisites to electronic organization.

The ultimaton is the first measurable form of energy, the prime physical unit of material existence. It is the particle below the electron, the foundation beneath the foundation.

Mutual attraction holds one hundred ultimatons together in the constitution of the electron; and there are never more nor less than one hundred ultimatons in a typical electron.

One hundred ultimatons compose a single electron. This is a precise claim, not a vague gesture toward subatomic complexity. Each electron in every atom in every substance in the universe is composed of exactly one hundred of these prime units, held together by mutual attraction in a specific configuration.

The loss of one or more ultimatons destroys typical electronic identity, thus bringing into existence one of the ten modified forms of the electron.

The electron is stable because its ultimatonic constitution is stable. Disturb that constitution, remove even one ultimaton, and you no longer have a normal electron but one of ten modified forms, a different entity with different properties.

No experiment has ever found evidence of subelectronic structure. The electron behaves, in every test we have devised, as if it were truly fundamental. If ultimatons exist, they are hidden below the threshold of our instruments, below the resolution of our most powerful particle colliders.

Should we dismiss the claim? Consider that physics has a long history of discovering structure where none was thought to exist. The atom was once considered indivisible; that is what the word means in Greek. Then we discovered electrons, protons, and neutrons. The proton was once considered fundamental. Then we discovered quarks.

What lies below quarks? The Standard Model does not say. String theory proposes that all particles are ultimately vibrating strings of energy, but string theory remains unconfirmed. The question of what constitutes the final, irreducible layer of physical reality, if such a layer exists, remains open. The text anticipates an answer:

> *There is innate in matter and present in universal space a form of energy not known on Earth. When this discovery is finally made, then will physicists feel that they have solved, almost at least, the mystery of matter.*

The ultimaton would be that layer. The place where energy first takes on the characteristics of matter. The point at which the great stream flowing from the cosmic center first crystallizes into something that can be counted, measured, and organized.

> *Ultimatons do not describe orbits or whirl about in circuits within the electrons, but they do spread or cluster in accordance with their axial revolutionary velocities, thus determining the differential electronic dimensions.*

The ultimaton does not orbit the way planets orbit stars or electrons orbit nuclei. It spreads and clusters. It has three varieties of motion: mutual resistance to cosmic force, individual revolutions of anti-gravity potential, and the intraelectronic positions of its hundred mutually interassociated companions within each electron.

These are not behaviors we can observe directly. We infer them from the description provided. But consider how strange the electron already is by the standards of everyday experience. It is not a tiny ball bearing orbiting a nucleus. It is a probability cloud, a smear of potential presence that only collapses into definite position when measured. Quantum mechanics has already taught us that the subatomic world operates by rules utterly foreign to our macroscopic intuitions.

The ultimaton would simply extend this strangeness one level deeper. A particle that responds to Paradise gravity but not to the linear gravity we measure in our laboratories. A particle that can accelerate to the point of partial antigravity behavior. A particle that, in nature, escapes the status of physical existence only when participating in the terminal disruption of a cooled-off and dying sun.

There are ten grand divisions of matter, a hierarchy stretching from the ultimatonic to the molecular. Let us trace this hierarchy upward, from the smallest to the most familiar.

At the base: ultimatonic matter, the prime physical units of material existence, the energy particles that compose electrons.

Next: subelectronic matter, the explosive and repellent stage of the solar supergases.

Then: electronic matter, the electrical stage of material differentiation —electrons, protons, and the various other units entering into the varied constitution of the electronic groups.

Beyond these: subatomic matter, found extensively in the interior of the hot suns.

Then: shattered atoms, found in the cooling suns and throughout space.

Next: ionized matter, individual atoms stripped of their outer electrons by electrical, thermal, or X-ray activities and by solvents.

Then: atomic matter, the chemical stage of elemental organization, the component units of molecular or visible matter.

Beyond atomic: molecular matter, matter as it exists on Earth in a state of relatively stable materialization under ordinary conditions.

Then: radioactive matter, the disorganizing tendency and activity of the heavier elements under conditions of moderate heat and diminished gravity pressure.

Finally: collapsed matter, the relatively stationary matter found in the interior of the cold or dead suns, matter subjected to unbelievable pressures.

This hierarchy, from the ultimatonic through the collapsed, describes the full range of material organization in the physical universe. Each level exhibits properties that could not be predicted from the level below. Each level represents a new stage of organization, a new set of behaviors, a new relationship between energy and form.

One of the boldest claims concerns the quantum nature of energy.

The relative integrity of matter is assured by the fact that energy can be absorbed or released only in those exact amounts which Earth scientists have designated quanta. This wise provision in the material realms serves to maintain the universes as going concerns.

This is precisely correct. And it is notable that this principle was stated so clearly in the early twentieth century.

In December 1900, Max Planck presented a paper to the German Physical Society that he himself didn't fully believe. He was trying to derive the correct formula for thermal radiation—the spectrum of light emitted by hot objects, which classical physics could not explain. The mathematics worked only if he assumed something that seemed absurd: that energy comes in discrete packets rather than continuous flows. Each packet, each quantum, has an energy proportional to its frequency.

Planck considered it a mathematical trick, a placeholder until someone found the real answer. He spent years trying to fit his discovery back into classical physics. He failed. Einstein extended the idea in 1905, proposing that light itself consists of discrete energy packets—what we now call photons—and used this to explain the photoelectric effect, work that earned him the Nobel Prize. Bohr applied quantization to atomic structure in 1913; the full framework of quantum mechanics wasn't established until the mid-1920s, when Heisenberg, Schrödinger, Born, and others developed the complete mathematical formalism. The text states quantization as settled cosmic fact during a period when physicists were still debating whether it was real.

What Planck had stumbled upon, almost against his will, was the foundation of quantum mechanics—the most successful and most strange of all physical theories. It tells us that at the smallest scales, nature is granular rather than smooth. Energy is not a fluid that can be divided infinitely; it is a currency that comes only in fixed denominations.

> *The quantity of energy taken in or given out when electronic or other positions are shifted is always a quantum or some multiple thereof.*

The cosmos deals in whole numbers, not fractions. In packets, not flows. The discrete nature of energy exchange is fundamental to the stability of matter itself.

There is one more claim worth considering: the proposition that space itself is not empty.

We think of the vacuum as nothingness, the absence of all substance. But physics tells us otherwise. Even the emptiest reaches of intergalactic space have measurable properties. Quantum field theory describes the vacuum as having a ground state with non-zero energy —effects like the Casimir force, where two uncharged metal plates placed very close together experience a tiny attraction because the vacuum between them has different properties than the vacuum outside.

Throughout all organized space there are gravity-responding energy currents, power circuits, and ultimatonic activities, as well as organizing electronic energies. Practically speaking, space is not empty.

Even far from any star or planet, the fabric of space contains substance.

The most nearly empty space known would yield about one hundred ultimatons, the equivalent of one electron, in each cubic inch. Such scarcity of matter is regarded as practically empty space.

One hundred ultimatons per cubic inch. The equivalent of one electron in a volume the size of a small die. This is as close to nothing as the physical universe gets, and still it is not nothing. The vacuum has a floor. Space has a minimum density. Even emptiness contains the seeds of matter.

What are we to make of this chapter's claims?

The ultimaton remains undetected, the hierarchy of matter unconfirmed by experiment. Contemporary physics has its own hierarchy—quarks, leptons, bosons—and its own open questions about what, if anything, lies beneath.

And yet.

The quantum nature of energy is correctly identified. The convertibility of matter and energy is correctly stated. The non-emptiness of space aligns with modern quantum field theory. Truths that physics was only beginning to articulate are grasped clearly. And beneath the physics lies a deeper principle:

In the evaluation and recognition of mind it should be remembered that the universe is neither mechanical nor magical; it is a creation of mind and a mechanism of law.

Perhaps the ultimaton will one day be discovered. Perhaps it is a placeholder for a reality that our instruments will eventually reveal. Or perhaps it is a way of gesturing toward the ineffable foundation of material existence, the place where energy first becomes countable.

What matters for our story is the vision: a universe built from the bottom up. Energy streaming from a cosmic source, crystallizing into particles, aggregating into atoms, combining into molecules, organizing into stars and planets and eventually into creatures capable of wondering about their own composition.

You are made of this material. The atoms in your body were forged in the hearts of ancient stars, scattered by supernova explosions, gathered into the dust cloud that became our solar system, incorporated into the planet that became your home, and eventually organized into the improbable structure that is reading these words.

From the cosmic center to the ultimaton to the electron to the atom to you: one unbroken chain of being, one continuous transformation of the primordial energy into ever more complex and conscious forms.

Without the Father would not anything exist that does exist.

The fabric of matter is also the fabric of mind. The story of particles is also, ultimately, the story of persons.[1]

3

BIRTH OF A NEBULA

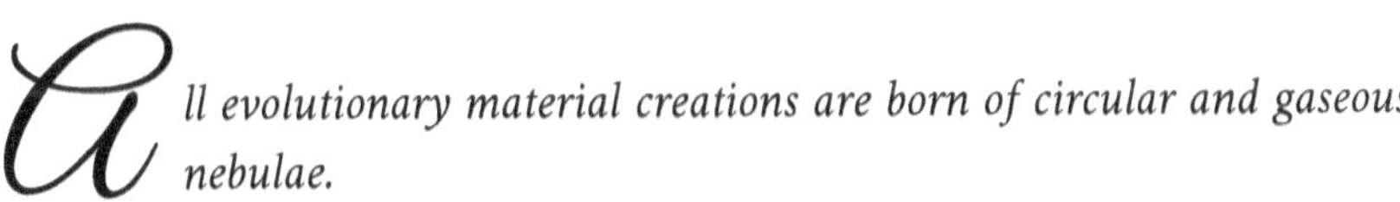

ll evolutionary material creations are born of circular and gaseous nebulae.

Now the story truly begins.

We have sketched the architecture of infinity, the cosmic center from which all things flow. We have descended into the fabric of matter, the particles that compose every star and stone. But these have been abstractions: grand patterns and tiny constituents, the very large and the very small. Now we narrow our focus to a single region of space, a particular cloud of gas and energy, a specific mother that will give birth to a specific family of suns, including the one that warms your face and grows your food and marks the passage of your days.

Earth is of origin in your sun, and your sun is one of the multifarious offspring of the Andronover nebula, which was onetime organized as a component part of the physical power and material matter of the local universe. And this great nebula itself took origin in the universal force-charge of space, long, long ago.

This is the story of the Andronover nebula: how it was born, how it grew, how it produced a million suns, and how it finally dispersed after giving all it had to give.

The universe, it seems, makes stars in order to make something else. What that something else might be, the nebula did not know. It simply spun, and contracted, and gave birth, and dispersed—playing its part in a story whose ending it would never see.

The numbers in this account are staggering, and we must address them honestly before we proceed. The Andronover nebula was initiated approximately 875 billion years ago. The first sun was born from it some 500 billion years ago. These numbers dwarf the timescales of conventional cosmology by orders of magnitude. Current astrophysics, based on observations of cosmic background radiation, the expansion rate of the universe, and the ages of the oldest stars, places the age of the observable universe at approximately 13.8 billion years. Not 875 billion. Not even close.

How are we to understand this discrepancy? Perhaps the framework operates with a different conception of "the universe," one that includes realms beyond our observable horizon, regions where time and space behave differently. Perhaps the numbers are symbolic rather than literal, expressing vastness rather than precise measurement. Or perhaps our own cosmological models are incomplete. The age of the universe is calculated from observations and assumptions that have been revised before and may be revised again. Dark energy, dark matter, the accelerating expansion of space: these are recent discoveries that have repeatedly surprised cosmologists. As we discussed in the first chapter, the Big Bang model faces serious problems that its proponents have not solved. The history of science counsels humility about what we think we know.

For our narrative purposes, we will present this account while acknowledging the tension with current scientific consensus. The reader may hold the numbers lightly, attending instead to the drama they describe: the birth of a stellar nursery, the slow whirl of gas into structure, the violent ejection of newborn suns. These processes are real. We observe them happening in nebulae throughout our galaxy today. Only the timeline is disputed.

Picture the scene. Before there was a nebula, there was only space. Not empty, as we discussed in the previous chapter, but thinly populated with the primordial energy-matter that pervades all of creation. Diffuse. Formless. Waiting.

Then something happened. The forces that organize matter throughout the cosmos arrived in this region of space. They came to initiate a new creation, to set in motion forces that would eventually produce stars and planets and, on at least one world, life itself.

> *875,000,000,000 years ago the enormous Andronover nebula number 876,926 was duly initiated. Only the presence of the force organizer and the liaison staff was required to inaugurate the energy whirl which eventually grew into this vast cyclone of space.*

An energy whirl. Imagine it: a region of space hundreds of light-years across, barely denser than the void around it, beginning to turn. So slowly at first that billions of years would pass before the motion became visible. But turning. The faintest spiral, the first hint of structure in the formless deep. Not an explosion, but a stirring. Not a bang, but a whirl. The forces that initiated this rotation then withdrew, their work complete.

> *Subsequent to the initiation of such nebular revolutions, the living force organizers simply withdraw at right angles to the plane of the revolutionary disk, and from that time forward, the inherent qualities of energy insure the progressive and orderly evolution of such a new physical system.*

Once set in motion, the nebula developed according to physical law. The initiators did not remain to micromanage its growth. They started the whirl and departed, trusting the inherent qualities of energy to carry the process forward. This is a vivid picture: not a cosmos that requires constant supernatural intervention, but one in which physical laws, once properly initiated, produce orderly results on their own.

Astrophysics today describes similar processes, though without the initial agency. Molecular clouds, vast regions of gas and dust in interstellar space, are stirred by various forces: the shock waves of nearby supernova explosions, the gravitational influence of passing stars, the density waves that create spiral arms in galaxies. These perturbations cause regions of the cloud to begin contracting under their own gravity, spinning faster as they shrink, flattening into disks. The mechanism differs, but the result is the same: a spinning disk of gas, the womb of future stars.

The Andronover nebula, once initiated, grew for hundreds of billions of years. It accumulated matter, drawing in gas and dust from the surrounding space. It rotated, flattened, and developed structure.

> *700,000,000,000 years ago the Andronover system was assuming gigantic proportions. At this distant date all of the material bequeathed to the subsequent creations was held within the confines of this gigantic space wheel, which continued ever to whirl and, after reaching its maximum of diameter, to whirl faster and faster as it continued to condense and contract.*

> *600,000,000,000 years ago the height of the Andronover energy-mobilization period was attained; the nebula had acquired its maximum of mass. At this time it was a gigantic circular gas cloud in shape somewhat like a flattened spheroid.*

At its maximum extent, the Andronover nebula spanned distances that dwarf our solar system the way our solar system dwarfs a grain of sand. Light itself would have taken years to cross from one edge to the other. It was one of the magnificent primary nebulae of this galactic region, a structure comparable to the entire star clouds we observe in the night sky—clouds that contain billions of stars, yet from Earth appear as faint smudges against the dark.

Within this vastness, matter was not uniformly distributed. Denser knots formed, regions where gravity began to overwhelm the outward pressure of heat. The seeds of future suns.

> *The near-by star students of that faraway era, as they observed this metamorphosis of the Andronover nebula, saw exactly what twentieth-century astronomers see when they turn their telescopes spaceward and view the present-age spiral nebulae of adjacent outer space.*

This is a striking observation. Astronomers studying Andronover in that distant epoch would have seen exactly what we see when we photograph the great spiral nebulae: arms of gas winding outward from a bright central core, regions of star formation glowing with the light of newborn suns. The processes are universal. What happened in Andronover billions of years ago is happening in nebulae throughout the cosmos right now.

> *About the time of the attainment of the maximum of mass, the gravity control of the gaseous content commenced to weaken, and there ensued the stage of gas escapement, the gas streaming forth as two gigantic and distinct arms, which took origin on opposite sides of the mother mass. The rapid revolutions of this enormous central core soon imparted a spiral appearance to these two projecting gas streams.*

The spiral arms that give spiral nebulae their characteristic beauty are not solid structures but patterns of density, regions where gas is compressed by the interplay of rotation and gravity. The cooling and subsequent condensation of portions of these arms produced their knotted appearance. These denser portions were vast systems and subsystems of physical matter whirling through space in the midst of the gaseous cloud, held securely within the gravity grasp of the mother nebula.

This matches observation. We see spiral structure throughout the universe, not just in nebulae but in galaxies, those vast congregations of billions of stars. The Milky Way itself is a barred spiral galaxy, its arms emerging from the ends of a central stellar bar and winding outward like a cosmic pinwheel. The physics of rotation naturally produces spiral forms: density waves move through the disk, compressing gas and triggering star formation, while differential rotation—matter moving at different speeds at different distances from the center—shapes the trailing arms.

And then the births began.

As the nebula contracted and its rotation accelerated, a critical threshold was reached. The centrifugal force of the outer regions began to exceed the gravitational pull holding them to the center. Material began to escape. Not randomly, but in structured ejections, blazing masses of gas thrown outward by the nebula's spin.

The energy wheel grew and grew until it attained its maximum of expansion, and then, when contraction set in, it whirled on faster and faster until, eventually, the critical centrifugal stage was reached and the great breakup began.

500,000,000,000 years ago the first Andronover sun was born. This blazing streak broke away from the mother gravity grasp and tore out into space on an independent adventure in the cosmos of creation.

The first sun. Born half a trillion years ago, according to this account, it broke free of the mother nebula and began its own journey through space. There is something almost joyful in this language. An adventure. The newborn star is not merely ejected; it is liberated. It begins its own story, even as it remains forever marked by its origin in the mother nebula.

Current stellar formation theory describes the process somewhat differently. Stars form, in our current understanding, not by centrifugal ejection from a spinning nebula, but by gravitational collapse within it. Dense knots of gas contract under their own weight, heating up as they compress, until the core becomes hot and dense enough to ignite nuclear fusion. The star turns on, begins to shine, and gradually disperses the remaining gas around it. The mechanism differs, but the poetry is the same: from diffuse cloud to blazing sun. From darkness to light. From potential to actuality.

Over the ages that followed, Andronover gave birth to more than a million suns. The precise number: 1,013,628 suns originated in this single nebula. Most were solitary orbs, like our own sun. Some were binaries, pairs of stars orbiting each other. Some were captured back into the nebular nucleus during phases of contraction; others escaped entirely, wandering off into the depths of space.

The process continued for hundreds of billions of years. The nebula went through cycles of contraction and expansion, each cycle producing new generations of stellar children. The oldest stars formed first, from gas closest to the nebular core. Younger generations formed later, from material in the spiral arms and outer regions.

Our sun was among the last children born. The 1,013,572nd sun to emerge from Andronover. The fifty-sixth from the last of its kind. A late child, born when the mother was already fading, already dispersing into the void.

> *And now the great Andronover nebula is no more. But it lives on in the many suns and their planetary families which originated in this mother cloud of space.*

There is genuine poignancy in this passage. A nebula is not a conscious entity; it does not experience its own dissolution. And yet the language invites us to feel something like grief, or at least recognition: the mother is gone, but the children remain. The substance that was Andronover now burns in a million stellar furnaces, including the one that lights our days.

We do not know where most of our sun's siblings are. Over billions of years, the stars born from Andronover have scattered across the galaxy, their original proximity lost to the differential rotation of the Milky Way's arms. Some may have died already. The most massive stars burn brightest but briefest, exhausting their fuel in mere millions of years and exploding as supernovae. Others, smaller and cooler than our sun, may still be burning, will still be burning long after our sun has swelled into a red giant and shed its outer layers into space.

Somewhere out there, other Andronover stars may host planets of their own. Other worlds may have developed life. Other minds may be contemplating their own origins, unaware that they share a common ancestor with us: a mother nebula that dispersed before any of them were born.

We are orphans, in a sense. But orphans with a million siblings, scattered across the stars.

Today, we can watch this process happening elsewhere.

The Orion Nebula, visible to the naked eye as a fuzzy patch in the sword of the constellation Orion, is a stellar nursery in active production. Within its glowing clouds, new stars are being born right now—or rather were being born 1,344 years ago, the time it takes light from that distance to reach us. The Hubble Space Telescope has captured images of protoplanetary disks within Orion: flattened rings of gas and dust surrounding young stars, the raw material from which planets will form.

The Eagle Nebula, home to the famous "Pillars of Creation" photographed by Hubble, shows columns of gas four to five light-years tall, within which new stars are condensing. The Carina Nebula, one of the largest and brightest in our sky, contains some of the most massive and luminous stars known, still embedded in the gas from which they formed.

These nebulae are not Andronover. Our mother cloud dispersed long ago. But they are her cousins, her contemporaries in the ongoing process of stellar formation that continues throughout the galaxy. By studying them, we glimpse the process that gave rise to our own sun and, eventually, to us.

The birth of a nebula is the beginning of a story. From diffuse energy comes structured matter. From formless gas comes spinning disk. From spinning disk come stars. From stars come planets. From planets comes life.

We have now set the stage. The Andronover nebula has been born, has grown, has begun to produce its children. In the next chapter, we will watch as one of those children, an ordinary yellow star undistinguished among its million siblings, ignites and begins its long, steady burn.

That star is our sun. And everything else in this story flows from its light.[1]

4

A STAR IS BORN

This blazing streak broke away from the mother gravity grasp and tore out into space on an independent adventure in the cosmos of creation.

Among the million suns born from the Andronover nebula, one concerns us above all others.

It was not the largest. It was not the brightest. It was not the first, nor the last, nor in any obvious way distinguished from its countless siblings. A medium-sized yellow star, modest by cosmic standards, unremarkable in its luminosity or mass. If you were a celestial cartographer cataloging the children of Andronover, you might pass over it without a second glance.

And yet this particular star, this one ordinary furnace of hydrogen fusion among a million others, would become the center of everything that matters to us. Around it would form a small family of planets. On one of those planets, conditions would align in just the right way. And from that alignment would emerge, over billions of years, the improbable phenomenon of life, and eventually the still more improbable phenomenon of minds capable of wondering about their own origins.

This is the story of our sun.

6,000,000,000 years ago marks the end of the terminal breakup and the birth of your sun, the fifty-sixth from the last of the Andronover second solar family.

Six billion years ago. Current measurements, based on radiometric dating of meteorites and analysis of solar evolution models, place the sun's age at approximately 4.6 billion years. The difference of roughly 1.4 billion years is significant but not enormous; the estimate is strikingly close for information compiled before radiometric dating was fully developed.

What matters more than the exact number is the sequence: the sun was among the last generation of stars born from Andronover. The great nebula was in its terminal phase, its final convulsions producing the last wave of stellar children before dispersing entirely.

This places our sun in context. We are not firstborn. We are not special. We are one among a million, distinguished only by the accident of our attachment: the fact that we happen to live on a planet orbiting this particular star and therefore care about its history in a way we do not care about the histories of its siblings.

Picture the birth. The Andronover nebula, by this late stage, had been producing suns for hundreds of billions of years. Its original circular form had long since evolved into a spiral structure, then partially collapsed back toward its center as gravity reasserted itself. Periodic eruptions sent waves of stellar material outward; periodic contractions drew some of that material back.

8,000,000,000 years ago the terrific terminal eruption began. Only the outer systems are safe at the time of such a cosmic upheaval. And this was the beginning of the end of the nebula. This final sun disgorgement extended over a period of almost two billion years.

7,000,000,000 years ago witnessed the height of the Andronover terminal breakup. This was the period of the birth of the larger terminal suns and the apex of the local physical disturbances.

In one of these final eruptions, a mass of gas separated from the nebular core. Spinning, condensing, heating as gravitational compression converted potential energy into thermal energy. The core grew denser. The temperature rose. At some point, the exact moment lost to time, the temperature and pressure at the center crossed a threshold. Hydrogen nuclei, stripped of their electrons and moving at tremendous speeds, began to collide with enough force to overcome their mutual electrical repulsion. They fused. Hydrogen became helium. Mass converted to energy according to Einstein's equation. Light poured outward.

A star was born. Our star.

5,000,000,000 years ago your sun was a comparatively isolated blazing orb, having gathered to itself most of the nearby circulating matter of space, remnants of the recent upheaval which attended its own birth.

The young sun was not the stable, steady presence we know today. It was still finding its equilibrium, still settling into the rhythm that would sustain it for billions of years.

Today, your sun has achieved relative stability, but its eleven and one-half year sunspot cycles betray that it was a variable star in its youth. In the early days of your sun the continued contraction and consequent gradual increase of temperature initiated tremendous convulsions on its surface. These titanic heaves required three and one-half days to complete a cycle of varying brightness. This variable state, this periodic pulsation, rendered your sun highly responsive to certain outside influences which were to be shortly encountered.

Variable stars are stars whose brightness fluctuates, sometimes dramatically, sometimes subtly, due to internal processes or external factors. Many young stars are indeed variable; they are still settling into the stable equilibrium that characterizes mature stellar life.

The sunspot cycle, the roughly eleven-year oscillation in solar activity that produces the dark spots visible on the sun's surface, is real and well-documented. Individual cycles vary considerably, ranging from about nine to fourteen years, with an average close to eleven years. Whether this cycle constitutes evidence of a more turbulent youth is speculative, but the image is evocative: a young sun, still finding its rhythm, its output fluctuating as it learned to burn steadily.

The three-and-one-half day pulsation cycle described in the text is intriguing. Young stars do exhibit various forms of variability. T Tauri stars, for example, are pre-main-sequence stars that show irregular brightness variations. Cepheid variables pulsate with periods ranging from days to months. Whether our sun went through such a phase is beyond our ability to confirm, but it is not implausible.

Our solar system's origin was unusual—less than one percent of planetary systems in this region of the galaxy formed the same way.

What made it so rare?

* * *

The young sun, still pulsating with the instabilities of youth, was not entirely alone. The space around it contained the usual debris: dust and gas left over from its formation, wandering chunks of rock and ice, the cosmic litter that accompanies any stellar birth. But something else was approaching. Something massive.

4,500,000,000 years ago the enormous dark system began its approach to the neighborhood of this solitary sun. The center of this great system was a dark giant of space, solid, highly charged, and possessing tremendous gravity pull.

A dark giant. Not a star but something else: a massive body, dark and cold, wandering through space on its own trajectory. No light announced its coming. It was visible only by the stars it blotted out as it passed, a moving absence against the backdrop of the galaxy. But its gravity was immense, and as it drew near our sun, our sun began to respond—its outer layers straining toward the intruder like tides toward a moon.

As the dark giant more closely approached the sun, at moments of maximum expansion during solar pulsations, streams of gaseous material were shot out into space as gigantic solar tongues. At first these flaming gas tongues would invariably fall back into the sun, but as the visitor drew nearer and nearer, the gravity pull of the gigantic stranger became so great that these tongues of gas would break off at certain points, the roots falling back into the sun while the outer sections would become detached to form independent bodies of matter, solar meteorites, which immediately started to revolve about the sun in elliptical orbits of their own.

This is a vivid picture of tidal disruption. The dark giant's gravity tugged at our sun's outer layers, drawing out tongues of superheated gas. The sun's own pulsations amplified the effect, sending material farther outward during each expansion. As the intruder drew closer, more and more of this material escaped the sun's immediate grasp.

This situation developed for about five hundred thousand years until the dark giant made its closest approach to the sun; whereupon the sun, in conjunction with one of its periodic internal convulsions, experienced a partial disruption; from opposite sides and simultaneously, enormous volumes of matter were disgorged. From the side facing the giant there was drawn out a vast column of solar gases, rather pointed at both ends and markedly bulging at the center, which became permanently detached from the immediate gravity control of the sun.

At the moment of closest approach, the sun essentially erupted. Material poured out from both sides: a vast column drawn toward the passing giant, and a corresponding ejection on the opposite side in tidal sympathy. The column of gas drawn toward the intruder was immense, bulging in the center, tapering at the ends, permanently separated from the sun.

This great column of solar gases which was thus separated from the sun subsequently evolved into the twelve planets of the solar system.

Twelve planets. Astronomy today recognizes eight, though some propose a ninth lurking in the outer reaches. The discrepancy may involve definitions, or bodies that have since been lost or reclassified. Regardless, the claim is that the planets formed not from a primordial disk surrounding the sun, but from material torn from the sun itself by the gravity of a passing giant.

This is the encounter hypothesis: planets formed from material torn from the sun by a passing object. It was seriously considered by astronomers in the early twentieth century—Chamberlin and Moulton proposed a version in 1905, Jeans and Jeffreys developed it further in the 1910s and 1920s. By mid-century, it had been largely abandoned in favor of the nebular hypothesis, which holds that planets form from the same disk of gas and dust that surrounds a young star.

The scientific consensus today strongly favors the nebular hypothesis. We observe protoplanetary disks around young stars throughout the galaxy. The physics of planet formation from such disks is well understood. The encounter hypothesis faces serious objections: close stellar passages are rare, and the material pulled from a star would more likely disperse than condense into planets.

Yet the text makes a specific claim: less than one percent of planetary systems formed this way. It does not assert that encounter formation is typical—only that it happened here. Our solar system, in this account, is genuinely unusual.

Is there any evidence for this? Some astronomers have noted that our solar system is atypical in various ways: the absence of "super-Earths" that are common elsewhere, the relatively circular orbits of our planets, the size and distance of our moon. Whether these oddities require an unusual formation mechanism remains debated. The text's claim cannot be confirmed, but it cannot be definitively ruled out either.

The five inner and five outer planets soon formed in miniature from the cooling and condensing nucleuses in the less massive and tapering ends of the gigantic gravity bulge which the dark giant had succeeded in detaching from the sun, while Saturn and Jupiter were formed from the more massive and bulging central portions.

Jupiter and Saturn, the giants of our solar system, formed from the densest part of the ejected column. The smaller planets formed from the thinner ends. This explains the distribution of mass in our solar system: the gas giants in the middle orbital zones, the smaller rocky and icy worlds closer in and farther out.

Jupiter and Saturn were, for a time, almost stars themselves. They glowed with their own light, radiating the heat of their formation. Jupiter does in fact radiate more heat than it receives from the sun—a residual glow from its violent birth. Had Jupiter been somewhat more massive, it might have ignited nuclear fusion and become a true binary companion to our sun.

The dark giant, meanwhile, continued on its way. It captured some of the ejected material for itself, three streams of solar substance that followed it into the depths of space. But most of the expelled matter remained bound to our sun, destined to become Mercury and Venus, Earth and Mars, Jupiter and Saturn and Uranus and Neptune, and all the smaller bodies that populate the spaces between.

All of the solar system material derived from the sun was originally endowed with a homogeneous direction of orbital swing, and had it not been for the intrusion of these three foreign space bodies, all solar system material would still maintain the same direction of orbital movement. As it was, the impact of the three captured tributaries injected new and foreign directional forces into the emerging solar system with the resultant appearance of retrograde motion.

This passage explains one of the peculiarities of our solar system: retrograde motion. Most objects in the solar system orbit the sun in the same direction, counterclockwise when viewed from above the sun's north pole. But some bodies, notably several moons of the outer planets, orbit in the opposite direction. The standard explanation is that these are captured objects. The text offers a different account: material captured by the dark giant from other systems, then gravitationally pulled into our solar system during the encounter, bringing with it foreign directional forces.

Retrograde motion in any astronomic system is always accidental and always appears as a result of the collisional impact of foreign space bodies. Such collisions may not always produce retrograde motion, but no retrograde ever appears except in a system containing masses which have diverse origins.

Retrograde motion as evidence of diverse origins. Our solar system, on this account, is not a pristine creation from a single source but a hybrid, incorporating material from our sun and from the system of the passing giant.

What kind of star is our sun?

Astronomers classify it as a G-type main-sequence star, a "yellow dwarf" in popular terminology, though it actually appears white from space; the yellow tint we see from Earth's surface is an artifact of atmospheric scattering. It is middle-aged, roughly halfway through its expected main-sequence lifetime of about ten billion years. It is metal-rich by astronomical standards, meaning it contains relatively high abundances of elements heavier than hydrogen and helium: carbon, oxygen, silicon, iron. The stuff of rocky planets and organic chemistry.

This metallicity is significant. The earliest stars, formed from primordial hydrogen and helium, contained almost no heavy elements. They could not have hosted planets like Earth. It was only after generations of stellar birth, death, and supernova enrichment that the interstellar medium became seeded with the heavier atoms necessary for rocky worlds and complex chemistry.

Our sun is a later-generation star, heir to the nuclear processing of countless predecessors. The iron in your blood, the calcium in your bones, the oxygen you are breathing: all of it was forged in stellar cores and scattered by stellar explosions before being gathered into the cloud that became our solar system.

You are, quite literally, made of star stuff. And not just any stars: the ancestors of our sun, the generations that preceded it, the cosmic lineage that stretches back through Andronover and beyond.

Somewhere out there, other Andronover stars may host planets of their own. Other worlds may have developed life. Other minds may be contemplating their own origins, unaware that they share a common ancestor with us: a mother nebula that dispersed before any of them were born.

Six billion years ago, or 4.6 billion by modern reckoning, a new light appeared in this arm of the galaxy. It was one light among many, one star among billions. It had no name then; names require namers, and the minds that would eventually name it "Sol" and "Helios" and "Sun" would not exist for billions of years yet.

But it burned. Steadily, faithfully, converting hydrogen to helium, mass to energy, matter to light. It would burn for billions of years more, is burning still as you read these words, casting its warmth across the small family of worlds that eventually formed around it.

On one of those worlds, that warmth would prove just right. Not too much, not too little. Just enough to permit liquid water, complex chemistry, the strange self-organizing processes we call life.

All of that lay in the future. For now, there was only the star: young, hot, blazing its light into the void, surrounded by a dispersing column of gas that would slowly condense into a family of worlds.

An independent adventure in the cosmos of creation had begun.[1]

5

———————

THE FORGING OF WORLDS

All of this tremendous activity is a normal part of the making of an evolutionary world and constitutes the astronomic preliminaries to the setting of the stage for the beginning of the physical evolution of such worlds of space in preparation for the life adventures of time.

In the beginning, fire and collision and cataclysm reigned on scales that beggar imagination.

The sun now burned in its corner of space, surrounded by a vast dispersing column of gas torn from its body by the gravitational violence of a passing giant. This was the raw material of worlds. But raw material is not yet a world. The transformation from gas cloud to planetary system would take billions of years and involve forces that would dwarf any violence in human experience.

What emerged from that chaos would be the solar system we know: four rocky inner worlds, four gas and ice giants in the outer reaches, a belt of asteroids marking the grave of a planet that never quite formed, and scattered debris extending to the frozen edges of the sun's gravitational domain. The architecture seems orderly now, almost inevitable.

Subsequent to the birth of the solar system a period of diminishing solar disgorgement ensued. Decreasingly, for another five hundred thousand years, the sun continued to pour forth diminishing volumes of matter into surrounding space. But during these early times of erratic orbits, when the surrounding bodies made their nearest approach to the sun, the solar parent was able to recapture a large portion of this meteoric material.

The young sun was not finished giving birth. For half a million years after the initial disruption, it continued to shed material into space, though in diminishing volumes. Some of this material fell back into the sun on close orbital passes. The rest joined the swirling chaos of gas and debris that would become the planets.

The planets nearest the sun were the first to have their revolutions slowed down by tidal friction. Mercury, closest to the solar furnace, experienced this braking effect most strongly. Today Mercury rotates slowly enough that its solar day—the time from one sunrise to the next—is 176 Earth days, actually longer than its 88-day year. The planet was once thought to be tidally locked, always showing the same face to the sun, but radar observations in 1965 revealed that Mercury is actually in a 3:2 spin-orbit resonance, rotating exactly three times for every two orbits.

Such gravitational influences also contribute to the stabilization of planetary orbits while acting as a brake on the rate of planetary-axial revolution, causing a planet to revolve ever slower until axial revolution ceases, leaving one hemisphere of the planet always turned toward the sun or larger body, as is illustrated by the planet Mercury and by the moon, which always turns the same face toward Earth.

Our moon is tidally locked to Earth, showing us the same face throughout its orbit. The same tidal forces that raise ocean tides on Earth have, over billions of years, slowed the moon's rotation until it matches its orbital period. Mercury, as we now know, did not quite reach full tidal lock but settled into a resonance that early observers mistook for synchronous rotation.

There is a prediction embedded in this cosmology that science has validated: the moon is slowly spiraling away from Earth, driven outward by tidal interactions. But this will not continue forever.

When the tidal frictions of the moon and the earth become equalized, the earth will always turn the same hemisphere toward the moon, and the day and month will be analogous, in length about forty-seven days. When such stability of orbits is attained, tidal frictions will go into reverse action, no longer driving the moon farther away from the earth but gradually drawing the satellite toward the planet.

Forty-seven day long days and months. The Earth and moon locked together, each showing the same face to the other perpetually. Modern physics supports the general outline: tidal evolution will eventually synchronize Earth's rotation with the moon's orbit, though this would take far longer than the sun's remaining lifespan. The specific numbers may differ, but the process is real. Whether tidal friction would then reverse, as the text claims, or whether some other mechanism—such as drag from the sun's expanding atmosphere— would draw the moon back toward Earth, remains speculative.

In that far-distant future when the moon approaches to within about eleven thousand miles of the earth, the gravity action of the latter will cause the moon to disrupt, and this tidal-gravity explosion will shatter the moon into small particles, which may assemble about the world as rings of matter resembling those of Saturn or may be gradu- ally drawn into the earth as meteors.

This is the Roche limit: the distance within which tidal forces will tear apart a body held together only by its own gravity. Any moon that spirals inside its planet's Roche limit will be shredded into rings. Saturn's magnificent rings may be the remains of just such a disrupted satellite. Our moon, in the very far future, may meet the same fate.

This passage describes a concept that Édouard Roche first calculated in 1848, though his work was not widely appreciated during his lifetime. Its accuracy is striking.

The great gas giants formed first and fastest.

The five inner and five outer planets soon formed in miniature from the cooling and condensing nucleuses in the less massive and tapering ends of the gigantic gravity bulge, while Saturn and Jupiter were formed from the more massive and bulging central portions.

Jupiter and Saturn, born from the densest part of the ejected column, were enormous from the beginning. Their powerful gravity captured most of the material that might otherwise have formed additional planets in their orbital zones. The asteroid belt, lying between Mars and Jupiter, is not a region where a planet failed to form; it is a region where Jupiter's gravity prevented one from forming.

Jupiter and Saturn, being derived from the very center of the enormous column of superheated solar gases, contained so much highly heated sun material that they shone with a brilliant light and emitted enormous volumes of heat; they were in reality secondary suns for a short period after their formation as separate space bodies.

Secondary suns. Jupiter and Saturn glowed with their own light in the early solar system, radiating the intense heat of their formation. They were not quite stars, lacking the mass to sustain nuclear fusion, but they were far more than cold, dark planets. Even today, Jupiter radiates more than twice as much energy as it receives from the sun, a remnant of that primordial heat still slowly leaking away after billions of years.

These two largest of the solar system planets have remained largely gaseous to this day, not even yet having cooled off to the point of complete condensation.

Jupiter and Saturn have no solid surfaces. Descending into their atmospheres, one would pass through ever-denser gas until the distinction between gas and liquid became meaningless, the pressure crushing any probe we could build. NASA's Juno mission has revealed that Jupiter's core is "fuzzy"—partially dissolved rock, ice, and metallic hydrogen with no clear boundary. No human will ever stand on the "ground" of Jupiter or Saturn. There is no ground to stand on.

4,000,000,000 years ago witnessed the organization of the Jupiter and Saturn systems much as observed today except for their moons, which continued to increase in size for several billions of years. In fact, all of the planets and satellites of the solar system are still growing as the result of continued meteoric captures.

The planets are still growing. Every meteor that falls to Earth adds mass to our world. Most are tiny, dust grains that burn up in the atmosphere. But occasionally, larger bodies arrive: the Chelyabinsk meteor that exploded over Russia in February 2013, the Tunguska event of June 1908, and in the deep past, the mountain-sized impactors that carved the great craters we can still see on the moon and other airless worlds.

Four billion years ago, the bombardment was relentless.

3,500,000,000 years ago the condensation nucleuses of the other ten planets were well formed, and the cores of most of the moons were intact, though some of the smaller satellites later united to make the present-day larger moons.

Ten planets. Astronomy recognizes eight, though some argue for a ninth in the distant outer reaches. The discrepancy may involve Pluto, now classified as a dwarf planet, or other bodies that have been lost or reclassified. The number matters less than the picture: by three and a half billion years ago, the basic architecture of the solar system was in place.

There is a story hidden in the asteroid belt.

The fifth planet of the solar system of long, long ago traversed an irregular orbit, periodically making closer and closer approach to Jupiter until it entered the critical zone of tidal disruption, was swiftly fragmentized, and became the present-day cluster of asteroids.

A fifth planet, orbiting between Mars and Jupiter, spiraling closer and closer to the giant until Jupiter's gravity tore it apart. The asteroid belt is its graveyard, the scattered fragments of a world that died before it ever truly lived.

Astronomy offers a different explanation: the asteroids are not the remains of a disrupted planet but primordial material that never coalesced into a planet in the first place, prevented from aggregating by Jupiter's gravitational influence. The total mass of the asteroid belt is only about three to four percent of the moon's mass, far too little to have ever been a substantial planet.

And yet, the cosmology presented here may preserve a kernel of truth. Recent studies suggest that the early solar system contained more planetary embryos than the planets we see today, and that gravitational interactions scattered or destroyed some of them. Models like the Nice model and the Grand Tack hypothesis incorporate planetary migration and possible embryo ejection. Perhaps a small planet did form in the asteroid belt and was later disrupted, even if the asteroids as a whole were never a single body.

Shooting stars occur in swarms because they are the fragments of larger bodies of matter which have been disrupted by tidal gravity exerted by nearby and still larger space bodies.

This is precisely correct. Meteor showers occur when Earth passes through the debris trails left by comets or other disrupted bodies. The Perseid meteor shower, every August, marks our passage through debris from Comet Swift-Tuttle. The fragments, spread along the comet's orbit, streak through our atmosphere year after year.

Saturn's rings are the fragments of a disrupted satellite.

Modern science agrees. Saturn's rings are believed to be the remains of a moon or moons that strayed within the planet's Roche limit and were torn apart by tidal forces. The rings are surprisingly young by astronomical standards, perhaps only 100 to 400 million years old based on Cassini mission data, suggesting that the disruption happened relatively recently in solar system history.

One of the moons of Jupiter is now approaching dangerously near the critical zone of tidal disruption and, within a few million years, will either be claimed by the planet or will undergo gravity-tidal disruption.

This likely refers to one of Jupiter's inner moons, probably Metis or Adrastea, which orbit within Jupiter's fluid Roche limit. They survive because they are held together not just by gravity but by their material strength. Eventually, tidal stresses may overcome this strength, and Jupiter too may gain a ring system from its moon's destruction.

3,000,000,000 years ago the solar system was functioning much as it does today. Its members continued to grow in size as space meteors continued to pour in upon the planets and their satellites at a prodigious rate.

Three billion years ago. The planets had settled into their orbits. The great bombardment was beginning to subside, though it would continue at diminishing intensity for hundreds of millions of years more. The solar system was becoming recognizable, though still violent by any standard we would consider comfortable.

About this time your solar system was placed on the physical registry of Nebadon and given its name, Monmatia.

Monmatia. The name of our solar system in the administrative records of the local universe. A designation that distinguishes it from the billions of other solar systems in this galactic region. Our address, in a sense, in the cosmic filing system.

2,500,000,000 years ago the planets had grown immensely in size. Earth was a well-developed sphere about one tenth its present mass and was still growing rapidly by meteoric accretion.

Earth at one-tenth its current mass. Still growing, still being built from the debris raining down from space. Every crater on the moon, every impact basin on Mars and Mercury, records this period of intense bombardment. Earth's craters have been erased by erosion and plate tectonics, but the evidence remains on our airless neighbors.

The solar system was taking shape. From chaos, order was emerging. The orbits were stabilizing. The planets were differentiating, heavy elements sinking to their cores, lighter materials rising to their surfaces. The stage was being set.

All of this tremendous activity is a normal part of the making of an evolutionary world on the order of Earth and constitutes the astro-nomic preliminaries to the setting of the stage for the beginning of the physical evolution of such worlds of space in preparation for the life adventures of time.

Normal. This violence, this bombardment, this chaos of forming worlds is normal. It is how planets are made throughout the universe. It is the astronomic preliminary to everything that matters to us: the emergence of stable worlds where chemistry can become biology, where matter can become mind.

The solar system was no longer a disaster zone. It was becoming a nursery.

In the next chapter, we will turn our attention to one world in particular: the third planet from the sun, a sphere of molten rock still being hammered by debris, still too hot to hold liquid water, still utterly lifeless. We will watch as that world cools, as its crust solidifies, as its oceans form.

We will watch the forging of Earth.[1]

6

WHEN EARTH WAS FIRE

he real geologic history of Earth begins with the cooling of the earth's crust sufficiently to cause the formation of the first ocean.

In the swirling disk of debris that surrounded our young sun, something was coalescing.

Not quickly. Nothing in planetary formation happens quickly by human standards. But inexorably. Particles of dust, drawn together by the faint gravity of their own accumulated mass, stuck to one another. Pebbles became rocks. Rocks became boulders. Boulders became mountains tumbling through space, sweeping up everything in their orbital paths.

This was the birth of Earth: not a single moment of creation, but a billion small additions. A world built grain by grain, stone by stone, impact by impact. The meteoric era of earth building.

Throughout these early times the space regions of the solar system were swarming with small disruptive and condensation bodies, and in the absence of a protective combustion atmosphere such space bodies crashed directly on the surface of Earth. These incessant impacts kept the surface of the planet more or less heated, and this, together with the increased action of gravity as the sphere grew larger, began to set in operation those influences which gradually caused the heavier elements, such as iron, to settle more and more toward the center of the planet.

Picture the early solar system, perhaps fifty million years after the sun's ignition. The protoplanetary disk has begun to clear. Much of the original gas has been blown away by solar radiation or swept up by the growing planets. What remains is a thinner but still substantial population of solid debris: chunks of rock and metal, ranging from dust grains to bodies hundreds of miles across.

In the zone that would become Earth's orbit, one object had grown larger than its neighbors. Perhaps it had a slight head start, a denser initial concentration of material, a favorable position in the disk. Whatever the cause, its greater mass gave it greater gravity, and greater gravity meant it swept up material faster than its competitors.

This is the runaway growth phase of planetary formation. The rich get richer. The largest body in each orbital zone grows at the expense of its smaller neighbors, vacuuming up debris, absorbing smaller planetesimals whole, gradually clearing its orbit of competition.

2,000,000,000 years ago the earth began decidedly to gain on the moon. Always had the planet been larger than its satellite, but there was not so much difference in size until about this time, when enormous space bodies were captured by the earth. Earth was then about one fifth its present size and had become large enough to hold the primitive atmosphere which had begun to appear as a result of the internal elemental contest between the heated interior and the cooling crust.

Earth's core, that dense ball of iron and nickel that still lies at our planet's center, formed during this phase. The heaviest elements, sinking through the molten interior of the growing planet, accumulated at the center. Lighter silicate rocks floated toward the surface.

The process is called differentiation: the separation of a planetary body into distinct layers based on density.

Definite volcanic action dates from these times. The internal heat of the earth continued to be augmented by the deeper and deeper burial of the radioactive or heavier elements brought in from space by the meteors.

There is a notable assertion embedded here: that radioactive dating will reveal Earth is more than one billion years old on its surface, but that such estimates are too short because the radioactive materials we can study are all derived from the earth's surface and hence represent comparatively recent additions. Radiometric dating places Earth's age at approximately 4.54 billion years, based primarily on analysis of meteorites, with the oldest terrestrial rocks and mineral grains providing corroborating evidence. The text's caution about surface materials representing recent acquisitions anticipates debates that would occupy geologists for decades.

The radium clock is your most reliable timepiece for making scientific estimates of the age of the planet.

Radium dating. Radiometric dating using uranium-lead decay chains, of which radium is an intermediate product, has indeed proven to be our most reliable method for dating ancient rocks. This endorsement of radiometric dating techniques, at a time when some religious traditions rejected such methods, is noteworthy.

The young Earth was not a hospitable place.

As the planet grew, the energy of each impact was converted to heat. Incoming planetesimals, striking at cosmic velocities, released tremendous energy upon collision. The planet's surface was not solid rock but liquid fire, a magma ocean that may have extended over a thousand miles deep.

1,500,000,000 years ago the earth was two thirds its present size, while the moon was nearing its present mass. Earth's rapid gain over the moon in size enabled it to begin the slow robbery of the little atmosphere which its satellite originally had.

Volcanic action is now at its height. The whole earth is a veritable fiery inferno, the surface resembling its earlier molten state before the heavier metals gravitated toward the center.

A veritable fiery inferno. The language is vivid and accurate. There were no gentle eruptions like Hawaii's flowing lava, no explosive blasts like Mount St. Helens. There was simply a world-spanning sea of liquid rock, churning and convecting, releasing heat to space, occasionally crusting over only to be shattered by another major impact.

The primitive planetary atmosphere is slowly evolving, now containing some water vapor, carbon monoxide, carbon dioxide, and hydrogen chloride, but there is little or no free nitrogen or free oxygen.

The atmosphere, such as it was, consisted of gases released from the planet's interior. Dense and hot and toxic, utterly unlike the air we breathe today. Current science suggests nitrogen was actually present in significant quantities alongside carbon dioxide, but free oxygen was indeed absent. If you could somehow transport yourself to that time, you would find no surface to stand on and no air to breathe. You would see a world that glowed with its own heat, red and orange and white, like an ember the size of a planet.

The atmosphere of a world in the volcanic age presents a queer spectacle. In addition to the gases enumerated it is heavily charged with numerous volcanic gases and, as the air belt matures, with the combustion products of the heavy meteoric showers which are constantly hurtling in upon the planetary surface. Such meteoric

This was the Hadean eon, named appropriately for Hades, the Greek underworld. It lasted from the formation of the solar system, approximately 4.57 billion years ago by current dating, until about 4 billion years ago. Over half a billion years of molten chaos.

———

Slowly, the bombardment lessened.

Not because the debris was exhausted. Plenty remained. But the largest bodies had swept their orbits relatively clean. The rate of major impacts declined. The intervals between catastrophic collisions grew longer.

And in those intervals, the Earth began to cool.

Cooling proceeded from the outside in. The surface, exposed to the cold of space, radiated heat away faster than it could be replenished from below. A skin began to form on the magma ocean, thin at first, fragile, repeatedly shattered by ongoing impacts, but each time reforming a little thicker, a little more stable.

The atmosphere, together with incessant moisture precipitation, facilitated the cooling of the earth's crust. Volcanic action early equalized internal-heat pressure and crustal contraction; and as volcanoes rapidly decreased, earthquakes made their appearance as this epoch of crustal cooling and adjustment progressed.

This was the first crust. Not the continents we know today—those would come much later—but a primitive skin of solidified rock, floating on the still-molten interior like ice on a winter pond. The crust was not uniform. Some regions cooled faster than others, creating variations in thickness and composition. The heavier basaltic rock that formed from the solidifying magma was dense; lighter minerals, segregating during cooling, would eventually form the granite that underlies our continents. But in the beginning, there were no continents, just a patchwork of cooling crust over a planetary ocean of liquid rock.

Then came the water.

Where did Earth's water originate? This remains one of planetary science's most debated questions. Some researchers argue that water was present in the rocky material that formed Earth originally, trapped in minerals, released by volcanic activity as the planet heated and differentiated. Others contend that most of Earth's water was delivered later, by comets and water-rich asteroids from the outer solar system. Most researchers now believe both mechanisms contributed.

As the crust cooled and stabilized, water began to accumulate on the surface. Volcanic outgassing released vast quantities of steam into the atmosphere. As temperatures dropped, this steam condensed and fell as rain—rain that would have flashed to steam again upon hitting the still-hot surface, rising, condensing, falling again in an endless cycle until the surface cooled enough for liquid to persist.

> *Presently, the atmosphere became more settled and cooled sufficiently to start precipitation of rain on the hot rocky surface of the planet. For thousands of years Earth was enveloped in one vast and continuous blanket of steam. And during these ages the sun never shone upon the earth's surface.*

Imagine that rain. Not a gentle drizzle but a deluge beyond comprehension, falling from a sky choked with volcanic gases, pooling in the low places of the newly solidified crust, steaming where it touched rock still warm from the planet's formation. Days of rain. Weeks. Years. Centuries. Millennia. A blanket of steam so thick and continuous that no sunlight penetrated to the surface.

When it was done, the Earth was wrapped in a global ocean.

Water-vapor condensation on the cooling surface of the earth, once begun, continued until it was virtually complete. By the end of this period the ocean was world-wide, covering the entire planet to an average depth of over one mile.

1,000,000,000 years ago is the date of the actual beginning of Earth history. The planet had attained approximately its present size.

One billion years ago marks a turning point. The planet had reached essentially its current mass. The great bombardment had largely subsided. The crust had stabilized enough to hold permanent oceans. The stage was set.

The real geologic history of Earth begins with the cooling of the earth's crust sufficiently to cause the formation of the first ocean.

There were no continents yet, or if there were, they were small scattered islands in a world-spanning sea. The water was not clear and blue but murky with dissolved minerals, hot from the planet's residual heat, chemically different from today's seawater.

Modern geology supports this picture of an early water world. Evidence from ancient zircon crystals, tiny minerals that survive from Earth's earliest epochs, suggests liquid water was present on Earth's surface by 4.4 billion years ago, only about 150 million years after the planet's formation. These crystals, found in rocks from Western Australia's Jack Hills, contain isotopic signatures that could only have formed in the presence of liquid water.

Earth was not slowly accumulating its oceans over billions of years. The water arrived early and in abundance. The planet was a water world almost from the beginning.

At the opening of this faraway era, Earth should be envisaged as a water-bound planet. Later on, deeper and hence denser lava flows came out upon the bottom of the present Pacific Ocean, and this part of the water-covered surface became considerably depressed. The first continental land mass emerged from the world ocean in compensatory adjustment of the equilibrium of the gradually thickening earth's crust.

The distinction between oceanic and continental crust began early. Oceanic crust, formed from dense basaltic rock, sits lower than the lighter continental masses—a relationship that persists today. The Pacific basin is characterized by this dense oceanic crust, though the ocean floor itself is geologically young, continuously recycled through plate tectonics. The ancient continental cratons, by contrast, preserve rocks dating back 3.6 to 4 billion years.

950,000,000 years ago Earth presents the picture of one great continent of land and one large body of water, the Pacific Ocean. Volcanoes are still widespread and earthquakes are both frequent and severe. Meteors continue to bombard the earth, but they are diminishing in both frequency and size. The atmosphere is clearing up, but the amount of carbon dioxide continues large. The earth's crust is gradually stabilizing.

One continent. One ocean. This is consistent with modern reconstructions of Earth's earliest continental configurations. Geologists call the earliest supercontinent Vaalbara, dating to approximately 3.6 billion years ago, though the evidence is fragmentary and contested. What is certain is that continental crust, lighter than oceanic crust, gradually accumulated and combined into larger masses over billions of years.

The continental land mass of this era increased until it covered almost ten percent of the earth's surface. Severe earthquakes did not begin until the continental mass of land emerged well above the water. When they once began, they increased in frequency and severity for

ages. For millions upon millions of years earthquakes have dimin-
ished, but Earth still has an average of fifteen daily.

Earthquake frequency varies dramatically by magnitude. The United States Geological Survey (USGS) records about 55 earthquakes per day of magnitude 4 or greater, though only about 15-16 major earthquakes (magnitude 7+) occur per year. The claim that earthquake frequency has diminished over geologic time is consistent with a cooling, stabilizing planet.

850,000,000 years ago the first real epoch of the stabilization of the earth's crust began. Most of the heavier metals had settled down toward the center of the globe; the cooling crust had ceased to cave in on such an extensive scale as in former ages. There was established a better balance between the land extrusion and the heavier ocean bed.

Beneath the ocean, the young Earth continued to evolve. The crust thickened. Volcanic activity, while still intense by modern standards, gradually diminished as the planet's interior cooled. The mantle, the thick layer of hot but solid rock between crust and core, began to convect in organized patterns, carrying heat from the interior to the surface.

This convection would eventually drive the movement of continents, the building of mountains, the recycling of oceanic crust into the deep Earth. But that story belongs to later chapters. For now, what mattered was stability: the Earth was becoming a world that could retain its ocean, regulate its temperature, and maintain the conditions that would eventually permit life.

From a cloud of gas and dust, a star had formed. Around that star, from debris torn by a passing giant, a world had coalesced. That world had differentiated into core and mantle and crust. It had cooled enough to hold liquid water. It had wrapped itself in a global ocean. The first continents were rising from the deep.

Earth was no longer just a planet. It was becoming a home.[1]

7

PREPARING THE CRADLE

On a planet where life has a marine origin, the ideal conditions for life implantation are provided by a large number of inland seas, by an extensive shore line of shallow waters and sheltered bays.

For hundreds of millions of years, Earth had been a world of fire and flood, of molten rock and crashing meteors, of steam and sulfur and chaos. Now, as the planet settled into middle age, it began to transform into something gentler: a world of water and sky, of shallow seas and emerging lands, of conditions increasingly favorable for something unprecedented in its history.

Life was coming. But first, the cradle had to be prepared.

600,000,000 years ago the commission of Life Carriers arrived on Earth and began the study of physical conditions preparatory to launching life on world number 606 of the local system.

Life on Earth was not an accident. Not a chance combination of chemicals in a primordial soup. It was a deliberate act of cosmic gardening. Beings with knowledge and purpose surveyed our young planet, assessed its readiness, and made plans for its biological future.

The account that follows is notable for its scientific precision about what those conditions actually were.

<hr>

The first requirement was water, specifically salt water.

> *The Life Carriers had projected a sodium chloride pattern of life; therefore no steps could be taken toward planting it until the ocean waters had become sufficiently briny. The Earth type of protoplasm can function only in a suitable salt solution.*

This turns out to be exactly right. All life on Earth is built on a sodium chloride foundation. The concentration of salt in human blood, approximately 0.9 percent, mirrors the concentration of the ancient seas in which our ancestors evolved. We carry the ocean inside us, quite literally. Every cell in your body floats in a solution whose chemistry echoes the primordial seas of six hundred million years ago.

> *All ancestral life, vegetable and animal, evolved in a salt-solution habitat. And even the more highly organized land animals could not continue to live did not this same essential salt solution circulate throughout their bodies in the blood stream which freely bathes, literally submerses, every tiny living cell in this "briny deep."*

> *Your primitive ancestors freely circulated about in the salty ocean. Today, this same oceanlike salty solution freely circulates about in your bodies, bathing each individual cell with a chemical liquid in all essentials comparable to the salt water which stimulated the first protoplasmic reactions of the first living cells to function on the planet.*

The fact that blood plasma and seawater share similar ionic compositions has been recognized since the late nineteenth century. René Quinton explored this connection in his 1897 work on seawater as an organic medium, and A.B. Macallum developed the idea further in the early twentieth century, arguing that blood plasma ionic concentrations represent an "heirloom from primeval sea life." But the deeper implication—that this similarity reflects our evolutionary heritage, that we are walking aquariums carrying the ocean within us—this understanding belongs to the twentieth century and beyond.

———

The second requirement was atmosphere.

Not the atmosphere we breathe today. That would come later, after billions of years of biological transformation. But an atmosphere nonetheless, one capable of filtering the deadly radiation of space while admitting the light and warmth that life requires.

> *The planetary atmosphere filters through to the earth about one two-billionth of the sun's total light emanation.*

One two-billionth. This tiny fraction sustains all life on Earth. The sun pours forth energy in quantities beyond human comprehension, and our thin shell of air captures just enough to warm our world without incinerating it.

> *The earth's atmosphere is all but opaque to much of the solar radiation at the extreme ultraviolet end of the spectrum. Most of these short wave lengths are absorbed by a layer of ozone which exists throughout a level about ten miles above the surface of the earth, and which extends spaceward for another ten miles.*

When these words were written, the stratospheric ozone layer had been discovered but its crucial importance for life was not widely appreciated. The text's emphasis on this protective shield, its location, its thickness, its function in screening ultraviolet radiation, reflects sophisticated understanding of atmospheric physics.

The ozone permeating this region, at conditions prevailing on the earth's surface, would make a layer only one tenth of an inch thick; nevertheless, this relatively small and apparently insignificant amount of ozone protects Earth inhabitants from the excess of these dangerous and destructive ultraviolet radiations present in sunlight.

About three millimeters—roughly one-tenth of an inch. If you could gather all the ozone in the atmosphere and compress it to sea-level pressure, it would form a layer thinner than a pencil. Yet this gossamer shield stands between all life on Earth and sterilizing ultraviolet death. When humans began destroying this layer with chlorofluorocarbons in the late twentieth century, the consequences became apparent within decades: increased skin cancers, damaged ecosystems, a reminder of how delicate the balance truly is.

But were this ozone layer just a trifle thicker, you would be deprived of the highly important and health-giving ultraviolet rays which now reach the earth's surface, and which are ancestral to one of the most essential of your vitamins.

Vitamin D. Synthesized in human skin through the action of ultraviolet light. The ozone layer must be thick enough to block the dangerous wavelengths yet thin enough to admit the beneficial ones. The margin is narrow. The balance is precise.

The third requirement was temperature.

Life as we know it operates within a narrow thermal band: roughly between the freezing and boiling points of water. Too cold, and biochemical reactions slow to a standstill. Too hot, and proteins denature, cell membranes dissolve, the delicate machinery of metabolism tears itself apart.

> *Were it not for the "blanketing" effect of the atmosphere at night, heat would be lost by radiation so rapidly that life would be impossible of maintenance except by artificial provision.*

The greenhouse effect. Before it became a byword for climate catastrophe, the greenhouse effect was simply a description of how atmospheres retain heat. Without greenhouse gases—carbon dioxide, water vapor, and others—Earth's average surface temperature would be well below freezing. Life depends on this thermal blanket.

The temperature profile of the atmosphere itself reveals the complexity of the life-support system we inhabit. From the surface upward, the text describes, temperature falls steadily for six or eight miles, reaching approximately seventy degrees below zero Fahrenheit. This is the troposphere, where weather happens. Above it lies the stratosphere, where temperature remains constant through forty miles of altitude. Higher still, temperatures climb again, reaching extraordinary values in the ionosphere where auroras dance.

> *The height of the earth's atmosphere is indicated by the highest auroral streamers, about four hundred miles.*

Four hundred miles. Beyond that lies the vacuum of space. Within that thin shell, protected from radiation by ozone, warmed by greenhouse gases, bathed in precisely filtered sunlight, life would find its home.

The fourth requirement was geography.

Life needed not just water but the right kind of water: shallow, warm, protected from the violence of the open ocean.

On a planet where life has a marine origin, the ideal conditions for life implantation are provided by a large number of inland seas, by an extensive shore line of shallow waters and sheltered bays; and just such a distribution of the earth's waters was rapidly developing.

These ancient inland seas were seldom over five or six hundred feet deep, and sunlight can penetrate ocean water for more than six hundred feet.

Modern marine biology confirms this crucial detail. Photosynthesis, the foundation of nearly all food chains, requires light. Light penetrates seawater only so far—the euphotic zone extends to roughly two hundred meters, or about six hundred fifty feet, though this varies with water clarity. Below that depth, the ocean is dark, dependent on organic matter drifting down from the sunlit zone above or on the rare chemosynthetic communities around hydrothermal vents. For life to flourish, to build the complex ecosystems that would eventually produce land plants and animals and humans, it needed shallow water.

Subsequently the commission of Life Carriers returned to their headquarters, preferring to await the further breakup of the continental land mass, which would afford still more inland seas and sheltered bays, before actually beginning life implantation.

The continents were drifting. This claim, stated matter-of-factly in a text written before continental drift was accepted by mainstream science, anticipates one of the twentieth century's great geological revolutions.

In 1912, a German meteorologist named Alfred Wegener proposed that the continents had once been joined and had slowly drifted apart. He assembled compelling evidence: the jigsaw fit of coastlines, matching fossils on opposite shores of the Atlantic, geological formations that continued from one continent to another. The scientific establishment ridiculed him. Without a mechanism to move continents through solid ocean floor, his theory seemed absurd.

Wegener died on the Greenland ice cap in 1930, leading an expedition to study polar weather. He was fifty years old. He never knew he was right. It would take another three decades before seafloor spreading and plate tectonics provided the mechanism he couldn't explain, transforming continental drift from heresy to orthodoxy.

The text we are drawing from was composed during those decades of rejection. Its authors stated continental drift as simple fact. They were right when the scientific consensus was wrong.

This does not prove the text's authority on other matters. But it does suggest that we should be cautious about dismissing claims simply because current scientific consensus rejects them. Consensus has been wrong before.

The continental land drift continued. The earth's core had become as dense and rigid as steel, being subjected to a pressure of almost 25,000 tons to the square inch, and owing to the enormous gravity pressure, it was and still is very hot in the deep interior. The temperature increases from the surface downward until at the center it is slightly above the surface temperature of the sun.

This alignment is more striking than it might appear. In the 1930s and 1940s, estimates of Earth's core temperature ranged widely, from 3,000 to over 10,000 degrees Celsius. The precise value depends on understanding the behavior of iron under extreme pressure—physics that wasn't fully worked out until the late twentieth century. Yet the text confidently states that Earth's core temperature "slightly exceeds" the sun's surface temperature, placing it in the range of 5,500-6,000 degrees Celsius. Modern measurements, using diamond anvil cells and shock wave experiments, have converged on approximately 5,700-6,000 degrees Celsius for the inner core, while the sun's surface (photosphere) measures about 5,500 degrees Celsius.

The text was right, within a narrow margin, about a value that geophysicists didn't nail down until decades later.

The outer one thousand miles of the earth's mass consists principally of different kinds of rock. Underneath are the denser and heavier metallic elements. Throughout the early and preatmospheric ages the world was so nearly fluid in its molten and highly heated state that the heavier metals sank deep into the interior. Those found near the surface today represent the exudate of ancient volcanoes, later and extensive lava flows, and the more recent meteoric deposits.

The mantle actually extends approximately 1,800 miles from surface to core, but the general picture is correct: a rocky outer layer overlying denser metallic elements in the core.

The geography of early Earth was nothing like today's familiar arrangement of continents and oceans. One vast landmass dominated the globe, already beginning to fracture along lines that would eventually produce the continental configuration we know.

Depression of the ocean bottom during the prelife ages had upthrust a solitary continental land mass to such a height that its lateral pressure tended to cause the eastern, western, and southern fringes to slide downhill, over the underlying semiviscous lava beds, into the waters of the surrounding Pacific Ocean.

The Pacific Ocean, that vast basin that covers more of Earth's surface than all the continents combined, was already the dominant feature of planetary geography. It was then, and remains today, the great counterweight to the continental masses, the deep depression into which the land tends to slide.

Even today the continents continue to float upon this noncrystallized cushiony sea of molten basalt. Were it not for this protective condition, the more severe earthquakes would literally shake the world to pieces. Earthquakes are caused by sliding and shifting of the solid outer crust and not by volcanoes.

This description of isostasy—the floating of continents on denser material beneath—and of earthquake mechanics reflects understanding that was still developing in the early twentieth century. The distinction between volcanic and tectonic earthquakes, clear here, would be clarified by the plate tectonic revolution of the 1960s.

The continental crust was stabilizing. The violent convulsions of Earth's youth were subsiding. The oceans were finding their basins. The atmosphere was clearing. The temperature was moderating.

The sea bottoms are more dense than the land masses, and this is what keeps the continents above water. When the sea bottoms are extruded above the sea level, they are found to consist largely of basalt, a form of lava considerably heavier than the granite of the land masses. Again, if the continents were not lighter than the ocean beds, gravity would draw the edges of the oceans up onto the land, but such phenomena are not observable.

One final consideration: the view from space.

Life on Earth would not exist in isolation from the cosmos. The planet orbits within a vast sea of radiation and particles, stellar winds and cosmic rays, the debris of exploding stars and the gentle but persistent pressure of sunlight itself.

During the earlier times of universe materialization the space regions are interspersed with vast hydrogen clouds, just such astronomic dust clusters as now characterize many regions throughout remote space.

The space environment matters. Our solar system moves through the galaxy, passing through regions of varying density, encountering clouds of gas and dust, exposed to the radiation of nearby stars. These cosmic influences affect climate, mutation rates, perhaps even the pace of evolution itself.

This short-ray energy charge of universe space is four hundred times greater than all other forms of radiant energy existing in the organized space domains.

Cosmic radiation. High-energy particles from distant sources—supernovae and active galactic nuclei and the mysterious processes that accelerate particles to relativistic velocities. This radiation penetrates our atmosphere, interacts with air molecules, produces cascades of secondary particles that reach the surface. It damages DNA, drives mutation, and has shaped the course of evolution since life began.

All of these essential cosmic conditions had to evolve to a favorable status before the Life Carriers could actually begin the establishment of life on Earth.

And so the stage was set.

The water was salty enough, the atmosphere protective enough, the shallow seas extensive enough, the cosmic environment stable enough. The continental landmass was fracturing, creating the sheltered bays and inland seas that would shelter the first experiments in living matter.

It should be made clear that Life Carriers cannot initiate life until a sphere is ripe for the inauguration of the evolutionary cycle. Neither can we provide for a more rapid life development than can be supported and accommodated by the physical progress of the planet.

The scientific content is striking. The salt concentration of seawater and blood. The ozone layer and ultraviolet protection. The greenhouse effect and temperature regulation. Continental drift and shallow seas. Core temperature and isostatic balance.

A cradle had been prepared. Life was about to begin.[1]

8

THE DAWN OF LIFE

e had planted the primitive form of marine life in the sheltered tropic bays of the central seas of the east-west cleavage of the breaking-up continental land mass.

And then there was life.

After billions of years of cosmic evolution, after the birth and death of countless stars, after the slow accretion of a planet from the debris of solar catastrophe, after the long cooling and the rain of millennia and the formation of the world-spanning ocean—in some warm shallow bay, in water lit by a younger sun, molecules began to copy themselves.

Nothing visible. Nothing dramatic. No lightning strike, no cosmic announcement. Just chemistry becoming something more than chemistry. A threshold crossed so quietly that no observer, had one existed, could have marked the moment. But something new had appeared in the universe. Something that would change everything.

550,000,000 years ago the Life Carrier corps returned to Earth. In co-operation with spiritual powers and superphysical forces we orga-

This is a provocative claim, stated without apology or hedging. Life did not arise spontaneously from chemistry. It was purposefully planted by beings with knowledge and purpose. The narrative is direct. The universe is not indifferent to life. Life is not an accident. It is, in this framework, a cosmic intention.

If the account is true, then life was not a chance occurrence but a deliberate act of creation. We are not accidents. We are intended—part of a plan that extends far beyond our world.

Science tells a different story about the timing, though not necessarily about the significance. The earliest evidence of life on Earth dates not to 550 million years ago but to approximately 3.5 billion years ago, when stromatolites—layered structures built by microbial communities—first appeared in the fossil record. Chemical signatures in ancient rocks suggest life may have existed even earlier, perhaps as far back as 4.1 billion years ago, only a few hundred million years after the planet formed.

The discrepancy is substantial. The timeline presented in the text places life's origin nearly three billion years later than current scientific evidence suggests. This is worth noting. The text's timeline is substantially at odds with scientific evidence.

This is not a minor discrepancy to be explained away. Three billion years is an enormous gap. If the text is accurate about the mechanisms of life's origin—deliberate implantation by celestial beings—then either our dating methods are catastrophically wrong, or the text is wrong about the timing.

The evidence for ancient life is substantial: stromatolites in Australian rocks dated to 3.5 billion years ago, chemical signatures of biological activity in rocks even older, a scientific consensus that has only strengthened over decades of research. If we accept radiometric dating as reliable—and the text itself endorses it as "your most reliable timepiece"—then the timeline discrepancy cannot be resolved.

Perhaps this is where readers must make their own peace with the material. The text offers insights that science has vindicated and claims that science contradicts. Both exist within the same pages. An honest engagement with this material requires holding both in mind.

Yet within its own chronological framework, the account is internally consistent and unusually detailed. The Life Carriers do not simply arrive and scatter microbes randomly. They survey conditions. They wait for the right moment. They choose specific locations for specific reasons.

All planetary life, aside from extraplanetary personalities, had its origin in our three original, identical, and simultaneous marine-life implantations. These three life implantations have been designated as: the central or Eurasian-African, the eastern or Australasian, and the western, embracing Greenland and the Americas.

Three implantations. Three separate locations. The same life, planted in three places at once. Why?

Our purpose in making three marine-life implantations was to insure that each great land mass would carry this life with it, in its warm-water seas, as the land subsequently separated. We foresaw that in the later era of the emergence of land life large oceans of water would separate these drifting continental land masses.

Foresight applied to biology. The Life Carriers knew the continents would drift apart. They planted life in three locations so that when the landmasses separated, each would carry its own population of living organisms, ensuring that life would spread across the entire planet rather than remaining confined to a single region that might be destroyed by geological catastrophe.

This is sophisticated reasoning. Continental drift was not accepted science when these words were written. Yet the authors understood that continents move, that oceans form between them, and that life's survival depends on geographic distribution.

That we are called Life Carriers should not confuse you. We can and do carry life to the planets, but we brought no life to Earth. Earth life is unique, original with the planet. This sphere is a life-modification world; all life appearing hereon was formulated by us right here on the planet; and there is no other world in all Satania, even in all Nebadon, that has a life existence just like that of Earth.

A clarification that complicates the picture. The Life Carriers did not import life from elsewhere. They created it here, using local materials, adapting their designs to local conditions. Earth's life is unique, found nowhere else in the cosmos.

The idea that life on Earth is genuinely unique—rather than seeded from some universal template—has implications for the search for extraterrestrial life. If the text is correct, we should not expect to find life elsewhere that resembles Earth's. Each world where life exists would have its own distinctive biology, adapted to its own conditions, created fresh rather than transplanted.

Astrobiology takes no position on this question. We have only one example of life to study: our own. Whether life elsewhere, if it exists, would be similar or radically different remains entirely unknown.

Earth is something special: a decimal planet.

In the local system there are only sixty-one worlds similar to Earth, life-modification planets. The majority of inhabited worlds are peopled in accordance with established techniques; on such spheres the Life Carriers are afforded little leeway in their plans for life implantation. But about one world in ten is designated as a decimal planet and assigned to the special registry of the Life Carriers; and on such planets we are permitted to undertake certain life experiments in an

One world in ten is designated for experimentation. Earth is such a
world. This framing has consequences for how we might interpret the
peculiarities of terrestrial biology: our DNA-based genetics, our
specific biochemistry, the particular path evolution has taken here.
These might be experimental features, tried on this world and
perhaps no other.

The phrase "life experiments" is evocative. It suggests that the development of life on Earth was not entirely predetermined, that there
was room for variation, for trying new approaches, for seeing what
would work. Evolution, in this framework, is not merely the blind
selection of random variations but a process unfolding within a
designed experimental context.

*500,000,000 years ago primitive marine vegetable life was well estab-
lished on Earth.*

The text reports a fifty-million-year gap between the initial
implantation and the establishment of primitive plant life. Life takes
time to gain a foothold. Even with deliberate introduction, the early
organisms must adapt, reproduce, spread, compete. Nothing happens
overnight on evolutionary timescales.

*Greenland and the arctic land mass, together with North and South
America, were beginning their long and slow westward drift. Africa
moved slightly south, creating an east and west trough, the Mediterranean basin, between itself and the mother body. Antarctica,
Australia, and the land indicated by the islands of the Pacific broke
away on the south and east and have drifted far away since that day.*

Continental drift, described matter-of-factly. The Americas moving west. Africa separating from the supercontinent. Australia and Antarctica beginning their long journey to their current positions. All of this described in a text written before plate tectonics became accepted science.

The geography of early Earth was nothing like today's familiar arrangement. A single great landmass dominated the globe, already fracturing along lines that would produce the modern continents. As these fragments drifted apart, they carried life with them, just as the Life Carriers had planned.

450,000,000 years ago the transition from vegetable to animal life occurred. This metamorphosis took place in the shallow waters of the sheltered tropic bays and lagoons of the extensive shore lines of the separating continents.

The transition from plant to animal. Modern biology understands that plants and animals diverged from distinct branches of the eukaryotic tree long ago—animals are actually more closely related to fungi than to plants. But the basic observation stands: at some point, organisms appeared that moved, that consumed other organisms, that exhibited the behaviors we associate with animals rather than plants.

And this development, all of which was inherent in the original life patterns, came about gradually. There were many transitional stages between the early primitive vegetable forms of life and the later well-defined animal organisms. Even today the transition slime molds persist, and they can hardly be classified either as plants or as animals.

Slime molds. A fascinating example, and one that remains relevant today. These organisms are now classified as protists, primarily within the supergroup Amoebozoa—neither plants nor animals nor fungi, though they were historically confused with all three. They exhibit characteristics superficially resembling plants (fruiting bodies), animals (movement, phagocytosis), and fungi (spore dispersal). Biologists now understand these similarities reflect convergent evolution rather than close relationship. The text uses them as evidence that life's categories are not always clear-cut.

> *Although the evolution of vegetable life can be traced into animal life, and though there have been found graduated series of plants and animals which progressively lead up from the most simple to the most complex and advanced organisms, you will not be able to find such connecting links between the great divisions of the animal kingdom nor between the highest of the prehuman animal types and the dawn men of the human races. These so-called "missing links" will forever remain missing, for the simple reason that they never existed.*

The material we are drawing from suggests that the major transitions in evolutionary history were not gradual accumulations of small changes but sudden appearances of new forms. The missing links that troubled Darwin and continue to generate debate in evolutionary biology are not missing because we haven't found them yet. They are missing because they never existed.

> *From era to era radically new species of animal life arise. They do not evolve as the result of the gradual accumulation of small variations; they appear as full-fledged and new orders of life, and they appear suddenly.*

Sudden appearance. This language echoes debates that would not become prominent in evolutionary biology until 1972, when Niles Eldredge and Stephen Jay Gould proposed the theory of punctuated equilibrium. Their theory suggested that evolutionary change is not constant but occurs in bursts, with long periods of stability interrupted by rapid transitions. The text appears to anticipate this pattern, though its explanation for the cause differs from the scientific one.

The sudden appearance of new species and diversified orders of living organisms is wholly biologic, strictly natural. There is nothing super-natural connected with these genetic mutations.

A clarification. The sudden appearances are not miraculous interventions but natural processes. The Life Carriers set up the conditions; the evolutionary mechanisms do the rest. This is a form of theistic evolution, but one that respects the natural order: God works through nature, not around it.

At the proper degree of saltiness in the oceans animal life evolved, and it was comparatively simple to allow the briny waters to circulate through the animal bodies of marine life.

Ocean salinity as an evolutionary factor. The text has already established that Earth life was designed around sodium chloride chemistry. Now it notes that the ocean's salt concentration enabled the development of animal life, which could use the surrounding seawater as a ready-made circulatory medium.

But when the oceans were contracted and the percentage of salt was greatly increased, these same animals evolved the ability to reduce the saltiness of their body fluids just as those organisms which learned to live in fresh water acquired the ability to maintain the proper degree of sodium chloride in their body fluids by ingenious techniques of salt conservation.

This describes osmoregulation, the ability of organisms to maintain internal salt balance regardless of their external environment. Marine fish must excrete excess salt through specialized gill cells; freshwater fish must actively conserve it. The evolution of these capabilities was essential for life to colonize diverse aquatic environments.

The physiologic equipment and the anatomic structure of all new orders of life are in response to the action of physical law, but the subsequent endowment of mind is a bestowal of the adjutant mind-spirits in accordance with innate brain capacity.

A distinction between body and mind. Physical evolution produces bodies according to natural law. Mind is something additional, bestowed rather than evolved, correlated with brain capacity but not identical to it. This dualism runs throughout the text's treatment of life, distinguishing the mechanisms of biological evolution from the spiritual dimensions of consciousness.

Mind, while not a physical evolution, is wholly dependent on the brain capacity afforded by purely physical and evolutionary developments.

Mind depends on a brain, even if mind is not reducible to a brain. The evolution of larger, more complex brains enables the bestowal of higher forms of mind. There is no mind without matter to support it, even if mind is more than matter.

Through almost endless cycles of gains and losses, adjustments and readjustments, all living organisms swing back and forth from age to age. Those that attain cosmic unity persist, while those that fall short of this goal cease to exist.

Life as a process of trial and success, gain and loss, adaptation and extinction. Most experiments fail. Most species that have ever existed are now extinct. Only those that achieve some form of stability—what the text calls "cosmic unity"—persist through the ages.

The dawn of life had broken over Earth. In warm shallow seas, in sheltered bays and tropical lagoons, something new was stirring. Simple at first, almost invisible, barely distinguishable from the chem-

istry of the waters around it. But alive. Growing. Changing. Beginning
the long journey that would eventually lead to trilobites and fish and
dinosaurs and mammals and, at last, to a particular species of primate
who would one day look back across the ages and wonder how it all
began.[1]

9

THE AGE OF THE SEAS

Suddenly and without gradation ancestry the first multicellular animals make their appearance. The trilobites have evolved, and for ages they dominate the seas. From the standpoint of marine life this is the trilobite age.

For nearly a quarter of planetary history, life remained confined to the oceans.

Think of what that means. From the first stirrings of living matter in warm shallow seas until the first tentative ventures onto land, hundreds of millions of years passed. Generation after generation of marine organisms lived and died and left their descendants in waters that lapped against barren, lifeless shores. The land waited, empty and silent, while the seas teemed with an ever-increasing diversity of life.

400,000,000 years ago marine life, both vegetable and animal, is fairly well distributed over the whole world. The world climate grows slightly warmer and becomes more equable. There is a general inundation of the seashores of the various continents, particularly of

North and South America. New oceans appear, and the older bodies of water are greatly enlarged.

This was the Paleozoic era, the age of ancient life, a span of nearly three hundred million years during which the seas served as the crucible of biological innovation. The Cambrian Explosion, early in this era, established almost all the major body plans that exist today within a geologically brief span of perhaps twenty-five million years. Every fundamental strategy for survival was tested in those ancient waters.

The first rulers of the seas were the trilobites.

But the trilobites were the dominant living creatures. They were sexed animals and existed in many forms; being poor swimmers, they sluggishly floated in the water or crawled along the sea bottoms, curling up in self-protection when attacked by their later appearing enemies.

Their bodies divided into three lobes, hence their name. Their shells articulated, allowing them to roll into protective balls when threatened. Their eyes, in some species, among the most sophisticated visual systems ever evolved: compound lenses made of crystalline calcite, some species boasting more than fifteen thousand individual facets.

They grew in length from two inches to one foot and developed into four distinct groups: carnivorous, herbivorous, omnivorous, and "mud eaters." The ability of the latter group largely to subsist on inorganic matter, being the last multicelled animal that could, explains their great increase and long survival.

The mud eaters. Organisms that could derive sustenance from the organic detritus accumulated in seafloor sediments. This ecological strategy proved successful. Trilobites that could process the endless rain of dead material falling from above had access to a food source that never ran out.

Three thousand varieties of brachiopods appeared at the close of this period, only two hundred of which have survived. These animals represent a variety of early life which has come down to the present time practically unchanged.

The fossil record reveals an even more dramatic story. Paleontologists have documented over twelve thousand brachiopod species across more than five thousand genera, with perhaps three hundred to four hundred species surviving today. The ratio of extinction remains staggering—roughly ninety-seven percent of all brachiopod species that ever existed are gone. What we see around us is not the full flowering of life's potential but the tattered remnants of ancient diversifications, the survivors of countless extinction events.

The trilobites are gone, but their legacy persists. The calcium carbonate of their shells became the limestone beneath our feet. We walk on the compressed remains of creatures who mastered the seas before anything with a backbone existed.

Brachiopods, those shelled creatures superficially resembling clams but built on an entirely different body plan, were among the dominant animals of Paleozoic seas. Today they are obscure, known mainly to specialists. But for hundreds of millions of years, they carpeted the seafloor in such abundance that their shells formed limestone layers hundreds of feet thick.

The seas were not empty before the trilobites. The text describes the earlier stages of marine life, before the appearance of complex animals.

The marine life was much alike the world over and consisted of the seaweeds, one-celled organisms, simple sponges, trilobites, and other crustaceans, shrimps, crabs, and lobsters.

This cosmopolitan distribution reflects a world without barriers. The continents had not yet fully separated. Ocean currents circulated freely around the globe. Life that evolved in one region could spread to others, creating a relatively uniform marine biosphere.

Lime-secreting algae were widespread. There existed thousands of species of the early ancestors of the corals. Sea worms were abundant, and there were many varieties of jellyfish which have since become extinct. Corals and the later types of sponges evolved.

Corals. These simple animals, building their calcium carbonate skeletons generation upon generation, would become the architects of some of Earth's most spectacular structures. The Great Barrier Reef, visible from space, is the work of billions of coral polyps over thousands of years. But the corals of the Paleozoic were different from modern reef builders, belonging to groups that are now entirely extinct.

The cephalopods were well developed, and they have survived as the modern pearly nautilus, octopus, cuttlefish, and squid.

Continuity across unimaginable spans of time. The nautiloid lineage originated in the Late Cambrian, approximately five hundred million years ago, though the modern genus Nautilus itself appeared much later, in the Triassic. Still, the basic body plan—the chambered shell, the tentacles, the jet propulsion—was perfected hundreds of millions of years ago. Evolution, having found a successful design, preserved it.

The trilobites rapidly declined, and the center of the stage was occupied by the larger mollusks, or cephalopods. These animals grew to be fifteen feet long and one foot in diameter and became masters of the seas. This species of animal appeared suddenly and assumed dominance of sea life.

Fifteen feet long. A foot in diameter. Imagine these giants, the terror of Ordovician seas, their tentacles grasping trilobites and smaller cephalopods, their parrot-like beaks crunching through shells, their jet-propelled bodies pursuing prey through waters that had never known such efficient predators. Recent reconstructions suggest some specimens of *Endoceras* may have reached nearly nineteen feet in length.

These were the nautiloids, straight-shelled relatives of today's coiled nautilus. They represent one of evolution's great experiments in predatory design. For tens of millions of years, they were the apex predators of the marine realm.

300,000,000 years ago another great period of land submergence began. The southward and northward encroachment of the ancient Silurian seas made ready to engulf most of Europe and North America.

The Silurian. Another period of the Paleozoic, another chapter in the long story of marine life. The continents rose and fell, the seas advanced and retreated, and life adapted to each new configuration of land and water.

The oceanic climate remained mild and uniform, and the warm seas bathed the shores of the polar lands. Brachiopod and other marine-life fossils may be found in these deposits right up to the North Pole.

Warm seas reaching toward the poles. The Silurian followed the severe end-Ordovician glaciation and represented a transition to greenhouse conditions. While the supercontinent Gondwana remained positioned over the South Pole, warming affected previously glaciated regions. For most of Earth's history—more than seventy percent of the past 4.5 billion years—the planet has maintained greenhouse conditions with minimal or no polar ice. The ice ages that have characterized the last few million years are the exception, not the rule.

*Gastropods, brachiopods, sponges, and reef-making corals continued
to increase.*

Gastropods. Snails and their relatives. These humble creatures, carrying their spiral homes on their backs, were among the most successful of Paleozoic marine animals. They crawled across seafloors, grazed on algae, preyed on other invertebrates, and diversified into thousands of species. Today, gastropods are the largest class of mollusks, with an estimated 65,000 to 80,000 living species.

*The close of this epoch witnesses the second advance of the Silurian
seas with another commingling of the waters of the southern and
northern oceans. The cephalopods dominate marine life, while associated forms of life progressively develop and differentiate.*

The oceans were not static. They expanded and contracted, connected and separated, rose and fell with the slow rhythms of continental movement and climate change. Each configuration created new opportunities for life, new niches to fill, new evolutionary experiments to run.

*280,000,000 years ago the continents had largely emerged from the
second Silurian inundation. The rock deposits of this submergence are
known in North America as Niagara limestone because this is the
stratum of rock over which Niagara Falls now flows.*

Niagara limestone. A connection between deep time and the present day that anyone can visit. Stand at Niagara Falls and you stand on rock formed from the accumulated shells of Silurian marine creatures. The thundering waters that have attracted tourists for centuries are carving through the compacted remains of animals that died over four hundred million years ago.

This layer of rock extends from the eastern mountains to the Mississippi valley region but not farther west except to the south.

The geography of ancient seas written in stone. Where limestone exists, seas once stood. Where it is absent, land persisted. The rock record is a map of vanished oceans, a chronicle of continental flooding and emergence preserved in layers that modern geologists can read like pages of a book.

But the greatest event of all was the sudden appearance of the fish family. This became the age of fishes, that period of the world's history characterized by the vertebrate type of animal.

The vertebrates. Our ancestors. For hundreds of millions of years, the seas had belonged to invertebrates: trilobites and brachiopods and cephalopods, creatures without internal skeletons, without back-bones, without the fundamental architecture that would eventually produce humans.

Then came the fish.

The marine life of this age was very diverse due to the early species segregation, but later on there was free commingling and association of all these different types. The brachiopods early reached their climax, being succeeded by the arthropods, and barnacles made their first appearance.

Succession. One group rises to dominance, flourishes, then declines as another takes its place. The brachiopods gave way to arthropods. The trilobites gave way to cephalopods. And all would eventually give way to the vertebrates, the backboned animals that would eventually crawl onto land and inherit the Earth.

The first fish were not the sleek, fast-swimming creatures we know today. They were armored, jawless, bottom-dwelling animals, more like living vacuum cleaners than the salmon and tuna of today's seas. They had no paired fins, no jaws, no teeth. They sucked organic matter from sediments, filtered food particles from water, lived slow lives in shallow coastal waters.

But they had backbones. They had the fundamental vertebrate body plan: a notochord, a dorsal nerve cord, pharyngeal slits, a post-anal tail. From this basic design, evolution would eventually produce sharks and bony fish, amphibians and reptiles, birds and mammals, and at last ourselves.

270,000,000 years ago the continents were all above water. In millions upon millions of years not so much land had been above water at one time; it was one of the greatest land-emergence epochs in all world history.

A world transformed. For the first time in hundreds of millions of years, the continents stood fully exposed. The shallow seas that had covered so much of the land retreated, leaving behind vast expanses of bare rock and sediment. On these newly exposed surfaces, the first land plants were already spreading, preparing the way for the colonization to come.

The marine life era was drawing to a close. Not in any absolute sense—life would remain abundant in the oceans to the present day. But its exclusive dominion was ending. The seas had served as the nursery of life, the laboratory where every fundamental body plan was tested, the arena where the great lineages that would dominate the future first appeared.

The rocks themselves record what was coming:

Some of the upper layers of these transition rock deposits contain small amounts of shale or slate of dark colors, indicating the presence of organic carbon and testifying to the existence of the ancestors of those forms of plant life which overran the earth during the succeeding Carboniferous or coal age.

But before coal could form, plants had to conquer the land. Before plants could conquer the land, they had to leave the water. And before they left the water, they had to evolve the structures that would allow them to survive in the harsh, desiccating environment of the terrestrial surface.

During these times of primitive marine life, extensive areas of the continental shores sank beneath the seas from a few feet to half a mile. Much of the older sandstone and conglomerates represents the sedimentary accumulations of these ancient shores.

Every sandstone cliff, every layer of shale, every bed of limestone tells a story. The sedimentary rocks of the world are the compressed remains of ancient environments: beaches and river deltas, shallow seas and deep ocean floors. Geologists reading these rocks can reconstruct vanished worlds, tracing the advance and retreat of seas, the rise and fall of mountains, the slow dance of continents across the face of the globe.

All of this story is graphically told within the fossil pages of the vast "stone book" of world record. And the pages of this gigantic biogeologic record unfailingly tell the truth if you but acquire skill in their interpretation.

The stone book. A poetic description of the geological record, but also an accurate one. The fossils embedded in sedimentary rocks are the written history of life on Earth, a record far more complete and reliable than any human chronicle. Learn to read it, and the entire history of the planet opens before you.

It is literally true, as your poet has said, "The dust we tread upon was once alive."

Every atom of calcium in your bones passed through the shells of countless marine creatures before reaching you. The limestone in your local buildings is the compacted remains of organisms that lived and died hundreds of millions of years ago. The chalk cliffs of Dover are made of the microscopic shells of single-celled algae that bloomed in Cretaceous seas. We walk on the dead. We build with the dead. We are ourselves temporary arrangements of atoms that have been part of life for billions of years and will be part of life for billions more.

The long age of the seas had established the fundamental patterns of life on Earth. Now the stage was set for the next great act: the conquest of the land.[1]

10

THE CONQUEST OF LAND

nd it was from such seashores of the mild and equable climes of a later age that primitive plant life found its way onto the land.

For three billion years, the land had waited.

Barren rock. Bare stone. Mountains that eroded without roots to hold their soil. Rivers that cut channels through landscapes devoid of any green thing. Coastlines where waves broke against shores that supported nothing more complex than the thin films of cyanobacteria that clung to wet surfaces at the tide line.

Then life climbed out of the water.

Vegetation now for the first time crawls out upon the land and soon makes considerable progress in adaptation to a nonmarine habitat.

This single sentence describes one of the most momentous transitions in the history of life. For hundreds of millions of years, all life had been aquatic. The land was as alien and hostile as the surface of Mars. No atmosphere to filter the sun's radiation. No moisture except what rain provided. Temperatures that swung wildly between scorching day and freezing night. Gravity that pressed down with a weight unknown to creatures buoyed by water.

Yet life made the crossing. Plants first, then animals. And in doing so, they transformed Earth from a blue planet with brown continents into a world of green.

There the high degree of carbon in the atmosphere afforded the new land varieties of life opportunity for speedy and luxuriant growth. Though this atmosphere was then ideal for plant growth, it contained such a high degree of carbon dioxide that no animal, much less man, could have lived on the face of the earth.

An atmosphere rich in carbon dioxide but low in oxygen. Favorable for plants, which consume carbon dioxide and release oxygen. Challenging for animals, which require oxygen to breathe. The colonization of land had to proceed in stages. Plants first, building the atmosphere that would eventually support animal life.

Reconstructions of Earth's atmospheric history confirm this sequence. Oxygen levels rose dramatically during the Silurian and Devonian periods, driven by the spread of land plants and the burial of organic carbon in sediments that would eventually become coal. The atmosphere we breathe today is the product of billions of years of biological activity, a manufactured environment created by life itself.

The oxygen you are breathing as you read this sentence was invented by those early land plants—not the exact molecules, but the process, the great engine of photosynthesis that turned a planet of carbon dioxide into a world where animals could live. Every breath connects you to the Silurian forests.

The first plants to colonize the land were humble things. Mosses and liverworts, lacking true roots or leaves, clinging to moist surfaces near streams and seeps. But these pioneers prepared the way. Their decomposing bodies created the first soils. Their photosynthesis began the slow oxygenation of the atmosphere. Their presence created microhabitats where more advanced plants could gain a foothold.

And it was from such seashores of the mild and equable climes of a later age that primitive plant life found its way onto the land. There the high degree of carbon in the atmosphere afforded the new land varieties of life opportunity for speedy and luxuriant growth.

Mild and equable climes. The colonization of land did not happen in harsh environments but in the gentle margins where land and sea met. Tidal pools and estuaries, river deltas and coastal swamps. Places where water was never far away, where the transition from aquatic to terrestrial could proceed by degrees.

Then came the great forests.

310,000,000 years ago begins the Carboniferous period, the age that would give us coal. This era witnessed one of the most dramatic transformations in Earth's history: the conversion of barren continents into vast swamp forests, dense with vegetation unlike anything that exists today.

The dominant plants of the Carboniferous were not trees in the modern sense. They were giant club mosses and horsetails, scale trees and tree ferns, plants whose living descendants are small and inconspicuous but whose ancestors grew to heights of a hundred feet or more. Lepidodendron, the scale tree, could reach 130 feet tall, its trunk marked with distinctive diamond-shaped leaf scars. Calamites, giant relatives of modern horsetails, formed dense stands along the margins of swamps. Tree ferns spread their fronds in the understory, thriving in the humid shade beneath the canopy.

But the world beneath these forests was restless:

> *The great volcanic activity of this age was in the European sector. Not in millions upon millions of years had such violent and extensive volcanic eruptions occurred as now took place around the Mediterranean trough and especially in the neighborhood of the British Isles. This lava flow over the British Isles region today appears as alternate layers of lava and rock 25,000 feet thick.*

The Carboniferous was not a peaceful time. Even as the forests flourished, volcanoes erupted, seas advanced and retreated, continents collided and mountain ranges rose. The volcanic sequences of the British Isles—including the Clyde Plateau Lavas in Scotland and later the Antrim Basalts—reach impressive thicknesses of several thousand feet and testify to this period of intense geological activity. The geology of this period laid down some of the most economically important rock formations on Earth: not just coal, but iron ore, limestone, and the mineral deposits that would later fuel industrial civilization.

> *The coal deposits of North America and Europe contain the organic remains of this age, the compressed and metamorphosed remnants of these primeval forests.*

Every lump of coal burned in a power plant or furnace was once living tissue. The carbon in that coal was once carbon dioxide in the Carboniferous atmosphere, captured by photosynthesis, built into plant bodies, buried in swamps, compressed by time and geology into solid fuel. When we burn coal, we are releasing energy stored by sunlight three hundred million years ago, returning carbon to the atmosphere that plants removed during the age of the great swamp forests.

The Carboniferous forests were unlike any ecosystem that exists today. The trees grew in waterlogged swamps where fallen trunks sank into anoxic muck, protected from decay, accumulating in layers that would eventually become coal seams hundreds of feet thick. The atmosphere was rich in oxygen, perhaps 25 to 35 percent compared to today's 21 percent. This elevated oxygen level had consequences.

Insects grew to extraordinary sizes.

Dragonflies with wingspans of two and a half feet. Millipedes over eight feet long—recent discoveries in northern England have revealed Arthropleura specimens reaching nearly nine feet in length. Scorpions the size of dogs. The Carboniferous was a world of giants, where creatures that today fit in the palm of your hand grew to dimensions that seem more appropriate to science fiction than natural history.

The oxygen-size connection is well established. Insects breathe through a network of tubes called tracheae, which deliver oxygen directly to their tissues. This system works well at small sizes but becomes inefficient as body size increases. Higher atmospheric oxygen levels allow the tracheal system to support larger bodies. The giant insects of the Carboniferous were possible because the atmosphere could support them.

The transition from vegetable to animal life occurred in the shallow waters of the sheltered tropic bays and lagoons.

But now that transition was repeating on land. Plants had made the crossing first. Invertebrates followed, tracking their food sources onto shore. Millipedes and scorpions, spiders and primitive insects, colonizing the new forests, adapting to air-breathing, learning to live without the support of water.

And then came the vertebrates.

But the greatest event of all was the sudden appearance of the fish family. This became the age of fishes, that period of the world's history characterized by the vertebrate type of animal.

Fish had dominated the seas for a hundred million years. Now some of them were developing features that would allow a new kind of life: fins that could support weight, lungs that could extract oxygen from air, skeletal structures that could function on land.

The earliest amphibians were not the delicate frogs and salamanders of today. They were robust creatures with powerful limbs and sturdy bodies adapted to hauling themselves across the mudflats of Devonian rivers. Ichthyostega, one of the first true tetrapods, reached about five feet in length. Later, during the Permian, amphibians would grow to truly massive sizes—Prionosuchus reached lengths of up to thirty feet—but the pioneers were more modest in scale.

In 2004, paleontologists searching for fossils in the Canadian Arctic, on Ellesmere Island, 600 miles from the North Pole, made a remarkable discovery. They had chosen this desolate location deliberately: rocks of exactly the right age, from exactly the period when fish should have been evolving toward land. After years of searching, they found what they were looking for.

They named it Tiktaalik, an Inuktitut word meaning "large freshwater fish." It was a snapshot of evolution in action: a creature with the scales and fins of a fish but with a neck that could turn, limbs that could prop up its body, and likely the capacity to breathe air. Neil Shubin, who co-led the discovery team with Edward Daeschler and Farish Jenkins Jr., later wrote that finding Tiktaalik was like "finding the missing link between fish and land animals." They had predicted where it would be, gone there, and found it.

Was this an accident? A fish that happened to survive a few gasping moments on a mudflat, passing that survival advantage to its descendants? Or was it, as the framework we are exploring suggests, a threshold that life was always meant to cross—the next step in a journey that began when the first cells stirred in warm Paleozoic seas?

In the seas, life continued to diversify even as the land invasion proceeded. But the arthropods, that ancient lineage of jointed-limbed creatures that includes insects, spiders, and crustaceans, were making the same transition that the vertebrates would follow. Scorpions and millipedes, ancient groups that had mastered air-breathing, were among the first complex animals to colonize the land, followed by the insects that would eventually dominate terrestrial ecosystems.

This metamorphosis took place in the shallow waters of the sheltered tropic bays and lagoons of the extensive shore lines of the separating continents. And this development, all of which was inherent in the original life patterns, came about gradually.

Gradually yet swift by geological standards. The transition from fully aquatic fish to fully terrestrial amphibians took perhaps twenty million years, a blink of an eye in the four-billion-year history of life. By the end of the Devonian, amphibians had established themselves on land, claiming a new ecological frontier.

But they were not truly free of the water. Amphibians must return to water or moist environments to reproduce. Their eggs lack shells; they must be laid in water or they will desiccate. Their skin is permeable; they lose moisture rapidly in dry air. They had conquered the land, but only partially. They remained tethered to the aquatic realm where their ancestors had evolved.

Throughout these times of climatic change, great variations also occurred in the land plants. The seed plants first appeared, and they afforded a better food supply for the subsequently increasing land animals.

The seed plants. Here was another revolution, less dramatic than the conquest of land itself but equally consequential for the future. Ferns and their relatives reproduce by spores, tiny cells that must find suitable moist conditions to germinate and grow. Seed plants package their embryos in protective coatings, providing food and shelter that allow them to survive drought, to disperse across distances, to colonize environments that spore plants cannot reach.

The evolution of seeds freed plants from dependence on water for reproduction, just as the evolution of the amniotic egg would later free vertebrates. Each innovation opened new possibilities, new environments, new evolutionary directions.

Gradually the inland lakes and seas were drying up all over the world. Isolated mountain and regional glaciers began to appear, especially over the Southern Hemisphere, and in many regions the glacial deposit of these local ice formations may be found even among some of the upper and later coal deposits. Two new climatic factors appeared: glaciation and aridity.

The world was changing. The warm, wet conditions of the Carboniferous were giving way to something harsher. Ice caps formed over the southern polar regions. Sea levels fell. The great swamp forests that had dominated the equatorial continents began to shrink, giving way to drier conditions that favored different kinds of plants, and different kinds of animals.

Many of the earth's higher regions had become arid and barren.

This was the Permian, the final period of the Paleozoic era, a time of environmental stress that would culminate in the greatest mass extinction in Earth's history. But before that catastrophe, the Permian would witness another great innovation: the rise of the reptiles, animals that had finally broken free from the water entirely.

160,000,000 years ago the land was largely covered with vegetation adapted to support land-animal life, and the atmosphere had become ideal for animal respiration.

The transformation was complete. What had been a world of barren rock and lifeless shores had become a world of forests and swamps, of insects and amphibians, of plants that reached for the sky and animals that crawled among their roots. The atmosphere, once challenging for animal life, had been converted by three hundred million years of photosynthesis into an oxygen-rich mixture capable of supporting active, warm-blooded metabolisms.

> *Thus ends the period of marine-life curtailment and those testing times of biologic adversity which eliminated all forms of life except such as had survival value, and which were therefore entitled to function as the ancestors of the more rapidly developing and highly differentiated life of the ensuing ages of planetary evolution.*

Testing times of biologic adversity. This phrase hints at what was coming: the Permian extinction, the Great Dying, an event that would eliminate more than eighty percent of marine species and perhaps seventy percent of terrestrial vertebrates. But the survivors would inherit a world transformed, a world where the conquest of land was complete, where the stage was set for the next great act in the drama of life.

> *The ending of this period of biologic tribulation, known to your students as the Permian, also marks the end of the long Paleozoic era, which covers one quarter of the planetary history, two hundred and fifty million years.*

The Paleozoic actually spanned closer to 290 million years, from the Cambrian explosion to the Permian catastrophe. During that time, life had gone from simple marine organisms to complex terrestrial ecosystems. It had invented the vertebrate body plan, the insect wing, the seed, the amniotic egg. It had transformed the atmosphere and covered the continents with vegetation.

> *The vast oceanic nursery of life on Earth has served its purpose. During the long ages when the land was unsuited to support life,*

before the atmosphere contained sufficient oxygen to sustain the higher land animals, the sea mothered and nurtured the developing life.

A poetic summary of what the oceans had been: the cradle of all living things, the protected environment where evolution could experiment without the harsh challenges of terrestrial existence. But now the nursery phase was over. Life had graduated to the land. The future belonged to creatures that could breathe air, walk on solid ground, and reproduce without returning to the water.

The conquest of land was complete. The age of reptiles was about to begin.[1]

11

THE REIGN OF REPTILES

hese rapidly evolving reptilian dinosaurs soon became the monarchs of this age.

For one hundred and fifty million years, reptiles ruled the Earth.

They dominated the land, filling every ecological niche from tiny insectivores to towering behemoths that shook the ground with their footsteps. They conquered the seas, evolving into streamlined predators that hunted fish and squid and one another in warm Mesozoic oceans. They took to the air, first with leathery wings stretched across elongated fingers, later with feathered bodies that would eventually give rise to birds.

This was the Mesozoic era, the time of the dinosaurs. And no other group of animals has so captured the human imagination. Museums display their mounted skeletons. Children memorize their names. Scientists debate their biology, their behavior, their ultimate fate. The dinosaurs are gone, yet they remain among the most famous creatures ever to walk the Earth.

120,000,000 years ago a new phase of the reptilian age began. The great event of this period was the evolution and decline of the dinosaurs. Land-animal life reached its greatest development, in point of size, and had virtually perished from the face of the earth by the end of this age.

The reptiles' great innovation was the amniotic egg.

Amphibians, those first vertebrate colonizers of land, remained tied to water. Their eggs, like those of fish, had no protective covering, no internal water supply, no mechanism for gas exchange with the atmosphere. They had to be laid in water or moist environments, where they could absorb oxygen through their permeable membranes and not desiccate in the dry air.

The amniotic egg made terrestrial reproduction possible. Inside its protective shell, the developing embryo floated in its own private pond, the amnion. A yolk sac provided nutrition. An allantois collected waste products. A chorion allowed gas exchange with the outside air. The reptilian egg was a self-contained survival capsule, allowing its makers to reproduce far from any body of water.

They were egg layers and are distinguished from all animals by their small brains, having brains weighing less than one pound to control bodies later weighing as much as forty tons.

One pound of brain for forty tons of body. This ratio seems absurd, yet it worked, at least for a time. The dinosaurs dominated Earth for longer than the entire history of mammals. Whatever their limitations, they were extraordinarily successful.

But earlier reptiles were smaller, carnivorous, and walked kangaroo-like on their hind legs. They had hollow avian bones and subsequently developed only three toes on their hind feet, and many of their fossil footprints have been mistaken for those of giant birds.

This observation presages one of the great discoveries of late twenti-eth-century paleontology: the connection between dinosaurs and birds. The hollow bones, the three-toed feet, the upright posture—all features that early observers noted as birdlike are now understood as evidence of direct ancestry. Birds did not merely resemble dinosaurs. Birds are dinosaurs, the only lineage of dinosaurs to survive the great extinction.

> *Later on, the herbivorous dinosaurs evolved. They walked on all fours, and one branch of this group developed a protective armor.*

The armored dinosaurs. Ankylosaurs with their bony plates and clubbed tails. Stegosaurs with their double row of plates and spiked tails. Nodosaurs covered in studded armor like living tanks. These creatures represent evolution's response to predation pressure, an arms race between hunters and hunted that produced some of the most bizarre body forms in Earth's history.

> *The dinosaurs evolved in all sizes from a species less than two feet long up to the huge noncarnivorous dinosaurs, seventy-five feet long, that have never since been equaled in bulk by any living creature.*

Seventy-five feet. Paleontology has found specimens even larger. Argentinosaurus may have reached one hundred feet in length. Patagotitan weighed perhaps fifty-five to sixty tons. These were the largest land animals that have ever existed, creatures of such immense scale that their very existence raises questions about biology and physics. How did hearts pump blood to brains thirty feet above ground? How did legs support such crushing weight? How did diges-tive systems process enough plant matter to fuel such enormous bodies?

> *The largest of the dinosaurs originated in western North America. These monstrous reptiles are buried throughout the Rocky Mountain regions, along the whole of the Atlantic coast of North America, over western Europe, South Africa, and India, but not in Australia.*

This geographic claim requires correction. Australia has in fact produced numerous dinosaur fossils, including Australotitan cooperensis—one of the world's ten largest dinosaurs at approximately thirty meters in length—along with Australovenator, Muttaburrasaurus, and Diamantinasaurus. The continent's dinosaur diversity was less than on other landmasses, but they were certainly present.

> *These massive creatures became less active and strong as they grew larger and larger; but they required such an enormous amount of food and the land was so overrun by them that they literally starved to death and became extinct. They lacked the intelligence to provide sufficient food to nourish such enormous bodies.*

This explanation for dinosaur extinction—that they were too large and too stupid to sustain themselves—was common in the early twentieth century but has since been superseded. We now know that the dinosaurs did not gradually decline due to their own inadequacies. They were thriving until the very end, cut down suddenly by a catastrophe: the asteroid impact at the end of the Cretaceous period.

This is one of the text's clear errors. The Chicxulub impact, discovered in the 1980s and now validated by multiple lines of evidence—the iridium layer, shocked quartz, the crater itself—explains the sudden extinction of the dinosaurs in a way that "growing too large" cannot. The dinosaurs did not gradually decline. They were eliminated in a geological instant.

Why would a text that correctly anticipated continental drift miss the asteroid impact? Continental drift had been proposed by Wegener in 1912, even if most geologists rejected it—the authors could affirm an existing hypothesis. But the asteroid impact wasn't even suggested until 1980; the Chicxulub crater wasn't confirmed until 1991. There was no hypothesis to affirm. The authors state they were restricted from revealing undiscovered science, and this would qualify.

Yet the underlying insight contains truth. The dinosaurs had evolved toward size rather than intelligence. Their brains were small relative to their bodies, their behavioral flexibility limited. When conditions changed catastrophically, they lacked the adaptability that smaller, smarter creatures possessed.

> *The dinosaurs, for all their enormous mass, were all but brainless animals, lacking the intelligence to provide sufficient food to nourish such enormous bodies. And so did these sluggish land reptiles perish in ever-increasing numbers. Henceforth, evolution will follow the growth of brains, not physical bulk, and the development of brains will characterize each succeeding epoch of animal evolution and planetary progress.*

This is the text's evolutionary lesson from the dinosaurs: that brains matter more than bulk, that intelligence will prove more valuable than size in the long run of life's history. The dinosaurs represent a path not taken, an alternative strategy of evolutionary success based on physical dominance rather than mental flexibility. Their extinction cleared the stage for a different kind of animal.

> *Soon after two species of dinosaurs migrated to the water in a futile attempt at self-preservation, two other types were driven to the air by the bitter competition of life on land.*

The seas and skies offered escape from the competitive pressure of the crowded land. Marine reptiles—ichthyosaurs and plesiosaurs and mosasaurs—returned to the water that their ancestors had left, evolving streamlined bodies and flippers and fishlike tails. Flying reptiles took to the air on wings of membrane stretched across elongated fourth fingers.

The marine reptiles deserve more attention than they often receive in popular accounts of the Mesozoic. Ichthyosaurs, with their dolphin-shaped bodies and enormous eyes, were among the most perfectly adapted marine predators ever to evolve. Plesiosaurs, with their long necks and paddle-like limbs, filled a different ecological role, perhaps feeding on fish and squid in ways analogous to modern seals. Mosasaurs, appearing later in the Cretaceous, grew to lengths of fifty feet, apex predators of the late Mesozoic seas.

But these flying pterosaurs were not the ancestors of the true birds of subsequent ages. They evolved from the hollow-boned leaping dinosaurs, and their wings were of batlike formation with a spread of twenty to twenty-five feet.

Twenty to twenty-five feet. Quetzalcoatlus, discovered in the 1970s, had a wingspan of over thirty feet, making it the largest flying animal ever known. These were not small creatures gliding tentatively through prehistoric skies. They were giants, aerial predators that soared over Mesozoic landscapes on wings no living creature can match.

These ancient flying reptiles grew to be ten feet long, and they had separable jaws much like those of modern snakes. For a time these flying reptiles appeared to be a success, but they failed to evolve along lines which would enable them to survive as air navigators. They represent the nonsurviving strains of bird ancestry.

The pterosaurs were an experiment in flight that ultimately failed. Their wing structure, based on membrane rather than feathers, may have been less efficient, less maneuverable, less adaptable than the feathered wings that birds would develop. Or perhaps they simply fell victim to the same catastrophe that eliminated the dinosaurs. In any case, they left no descendants.

One hundred million years ago the reptilian age was drawing to a close. The great Cretaceous period derives its name from the predominance of the prolific chalk-making foraminifers in the seas. This period brings Earth to near the end of the long reptilian dominance

*and witnesses the appearance of flowering plants and bird life on
land.*

The flowering plants. The angiosperms. Perhaps the most significant
evolutionary development of the Cretaceous, rivaling even the
appearance of birds in its long-term consequences. Before the
angiosperms, land plants reproduced by spores or by naked seeds
borne on cones. The flowering plants enclosed their seeds in protective structures, enlisted insects and other animals as pollinators, and
developed fruits that encouraged seed dispersal. They transformed
terrestrial ecosystems, creating the colorful, fragrant, fruit-bearing
world that we know today.

Another transformation was taking shape in the skies:

> *55,000,000 years ago the evolutionary march was marked by the
> sudden appearance of the first of the true birds, a small pigeonlike
> creature which was the ancestor of all bird life. This was the third
> type of flying creature to appear on earth, and it sprang directly from
> the reptilian group, not from the contemporary flying dinosaurs nor
> from the earlier types of toothed land birds.*

Current paleontology has revised this picture. Birds did not evolve as
a third, separate experiment in flight. They evolved from small feathered dinosaurs, theropods related to Velociraptor and Deinonychus,
creatures that were already warm-blooded, already covered in feathers, already possessing the light hollow bones that would make flight
possible.

In 1861, workers in a limestone quarry in Bavaria, Germany, split
open a slab of rock and found something impossible: a skeleton with
feathered wings, preserved in exquisite detail, but also with teeth,
clawed fingers, and a long bony tail. It was announced just two years
after Darwin published *On the Origin of Species*—a transitional form,
exactly the kind of evidence Darwin's critics claimed didn't exist.

They named it Archaeopteryx, "ancient wing." It remains one of the most important fossils ever discovered: a creature caught in the act of becoming something new.

The birds are not a separate creation but the living continuation of the dinosaur lineage. Every sparrow outside your window, every eagle soaring overhead, and every chicken in a farmyard is a dinosaur, the sole surviving branch of a family that once ruled the world.

And so this becomes known as the age of birds as well as the declining age of reptiles.

Several million years later the first mammals appeared. They were nonplacental and proved a speedy failure; none survived. This was an experimental effort to improve mammalian types, but it did not succeed on Earth.

This statement requires clarification. Mammals actually appeared much earlier than the text suggests, in the Triassic period, coexisting with dinosaurs for approximately 140 to 160 million years. But they remained small, inconspicuous, mostly nocturnal, relegated to the margins of ecosystems dominated by reptiles. The early experiments in mammalian body plans, including the non-placental types mentioned here, did mostly fail. Only three groups survived: the monotremes (egg-laying mammals like the platypus), the marsupials (pouched mammals like kangaroos), and the placentals (which include most modern mammals).

This period, embracing the height and the beginning decline of the reptiles, extended nearly twenty-five million years and is known as the Jurassic.

The Jurassic. The period that gave its name to the most famous dinosaur movie in history. But the Jurassic was only the middle act of the reptilian drama. The Triassic came before, witnessing the first appearance of dinosaurs. The Cretaceous came after, bringing the dinosaurs to their greatest diversity before their sudden end.

The great Cretaceous period was drawing to a close, and its termination marks the end of the great sea invasions of the continents.

The end of the Cretaceous. Sixty-six million years ago. An asteroid six miles wide struck what is now the Yucatan Peninsula of Mexico, releasing energy equivalent to billions of nuclear weapons. Tsunamis swept across oceans. Fires burned across continents. Dust and debris blocked sunlight for months or years. The photosynthetic base of food chains collapsed. The dinosaurs, along with the pterosaurs, the marine reptiles, the ammonites, and countless other groups, vanished from Earth.

In the late 1970s, Walter Alvarez was studying limestone formations in a gorge near Gubbio, Italy, trying to understand the rate at which ancient sediments accumulated. He noticed a thin layer of clay at the boundary between Cretaceous and Paleogene rocks—the exact moment when the dinosaurs disappeared. Working with his father Luis, a Nobel Prize-winning physicist, along with nuclear chemists Frank Asaro and Helen Michel, the team analyzed the clay and found it was enriched with iridium, an element rare on Earth but common in asteroids.

In 1980, they published their hypothesis: an asteroid impact had killed the dinosaurs. The scientific community was skeptical, even hostile. The crater had actually been detected in 1978 by geophysicist Glen Penfield while surveying for an oil company, but his findings received little attention. It took until 1991 for the scientific paper by Hildebrand and colleagues to formally confirm the crater, buried beneath the Yucatan Peninsula at Chicxulub, Mexico. The evidence has only strengthened since: shocked quartz, spherules of melted rock, tsunami deposits ringing the Gulf of Mexico. The Alvarez team was right. The dinosaurs died in a cataclysm of cosmic violence.

The text was written decades before any of this was known. But the result was the same, whether attributed to gradual decline or sudden catastrophe. The age of reptiles ended. The age of mammals was about to begin. The small furry creatures that had lived in the shadows of dinosaurs for over 150 million years would inherit a devastated world and remake it in their own image.

If the cosmos has purposes, they are not always gentle. The dinosaurs reigned for 150 million years—far longer than mammals have yet existed. Their extinction opened the door for creatures like us. Whether this was cosmic intention or cosmic accident, the result was the same: the path to human consciousness required their removal.[1]

1 2

THE AGE OF MAMMALS

*N*o animal the size of an elephant could have survived unless it had possessed a brain of large size and superior quality. In intelligence and adaptation the elephant is approached only by the horse and is surpassed only by man himself.

When the dust settled after the great extinction, the world belonged to the mammals.

They had waited 150 million years for this moment. Small, furtive, nocturnal creatures hiding in the underbrush while dinosaurs ruled the daylight hours. Now their moment had come. Within ten million years of the extinction, mammals had diversified to fill every ecological niche the dinosaurs had vacated. Within twenty million years, they had produced forms more varied and more specialized than the reptiles had ever achieved.

In North America the placental type of mammals suddenly appear, and they constitute the most important evolutionary development up to this time. Previous orders of nonplacental mammals have existed,

The placental mammals. Creatures that nurture their young inside the mother's body, connected by a placenta that provides nutrition and removes waste, allowing development to proceed to a much more advanced stage before birth. This reproductive strategy, demanding more of mothers but producing more capable offspring, would prove extraordinarily successful.

The father of the placental mammals is a small, highly active, carnivorous, springing type of dinosaur.

This suggests that mammals arose from dinosaurs rather than from an earlier reptilian lineage. Modern paleontology tells a different story: mammals evolved from synapsids—often misleadingly called "mammal-like reptiles"—specifically from a therapsid subgroup called cynodonts. The synapsid lineage diverged from the sauropsid lineage (which would eventually produce dinosaurs) approximately 315 million years ago, long before the first dinosaur appeared. Mammals and dinosaurs shared a common amniote ancestor but evolved along completely separate paths. Still, the spirit of the claim—that mammals arose from small, active ancestors trending toward warm-bloodedness—is not far from current understanding.

45,000,000 years ago the continental backbones were elevated in association with a very general sinking of the coast lines. Mammalian life was evolving rapidly.

The Eocene epoch. A greenhouse world, far warmer than today, with tropical forests extending into polar latitudes, with palm trees growing in Wyoming and crocodiles basking on the shores of the

Arctic Ocean. In this warm, wet world, mammals diversified at an astonishing rate.

A small reptilian, egg-laying type of mammal flourished, and the ancestors of the later kangaroos roamed Australia. Soon there were small horses, fleet-footed rhinoceroses, tapirs with proboscises, primitive pigs, squirrels, lemurs, opossums, and several tribes of monkeylike animals.

The catalog of early Cenozoic mammals reads like a roll call of familiar forms, yet all were primitive versions of their modern descendants. The horses were the size of dogs, with multiple toes instead of hooves. The rhinoceroses were nimble runners, not the ponderous beasts we know today. The monkeylike animals were the first stirrings of the primate lineage that would eventually produce humans.

They were all small, primitive, and best suited to living among the forests of the mountain regions.

Small and forest-dwelling. This was the ancestral condition for most mammalian lineages. The open grasslands that would later dominate much of Earth's land surface did not yet exist. Mammals lived in forests, climbing trees, foraging on the forest floor, adapting to a three-dimensional world of branches and canopy and undergrowth.

The mammals of the early Cenozoic lived on land, under the water, in the air, and among the treetops.

Every environment. Whales and their relatives returned to the sea, reversing the transition that their distant ancestors had made hundreds of millions of years before. Bats took to the air, the only mammals to achieve true powered flight. Primates colonized the canopy, developing the grasping hands and binocular vision that would later prove so useful for tool use and three-dimensional thinking.

The return to the sea. Whales evolved from hoofed land mammals, relatives of modern hippos, in a transition well documented by fossils from Pakistan and India. Within approximately ten to eleven million years, four-legged land animals had transformed into fully aquatic creatures, their front limbs becoming flippers, their hind limbs disappearing, their bodies becoming streamlined for life in the water. It is one of the most dramatic evolutionary transitions in the fossil record.

35,000,000 years ago marks the beginning of the age of placental-mammalian world domination.

The Oligocene epoch, which began approximately 34 million years ago. The world was cooling, the tropical forests retreating, the first grasslands beginning to spread across the continental interiors. This environmental shift would reshape mammalian evolution, favoring new body forms adapted to life on the open plains.

Formerly the mammals have lived for the greater part in the hills, being of the mountainous types; suddenly evolution of the plains or hoofed type, the grazing species, begins. These grazers spring from an undifferentiated ancestor having five toes and forty-four teeth, which perished before the end of the age.

The grazers. Horses, cattle, antelope, deer—all descended from forest-dwelling ancestors that ventured onto the spreading grasslands and found opportunity there. The five-toed ancestor gave way to forms with fewer toes, eventually producing the single-toed horses that could run faster than any predator.

20,000,000 years ago was indeed the golden age of mammals.

The Miocene. An epoch of extraordinary mammalian diversity, when the grasslands had spread across the continents and the animals that grazed upon them had diversified into hundreds of species. This was the world that would have seemed familiar to a present-day observer: savannas and prairies, herds of grazing animals, predators stalking them through tall grass.

The Bering Strait land bridge is up, and many groups of animals migrate to North America from Asia, including the four-tusked mastodons, short-legged rhinoceroses, and many varieties of the cat family.

The land bridges. Throughout the Cenozoic, the continents were connected and disconnected by land bridges that rose and fell with changing sea levels. These bridges served as highways for migration, allowing animals to spread from one continent to another, mixing faunas that had evolved in isolation. The Bering land bridge between Asia and North America was among the most important, channeling successive waves of migration in both directions.

The first deer appeared, and North America was soon overrun by ruminants: deer, oxen, camels, bison, and several species of rhinoceroses.

North America became a center of mammalian evolution. Many groups that we now associate with other continents—including horses and camels—actually originated in North America and migrated outward. The horses spread to Asia and Africa, then went extinct in their homeland around 10,000 years ago, only to be reintroduced by Spanish conquistadors in the sixteenth century. The camels spread to Asia (where they became the Bactrian camels) and South

America (where they became the llamas and alpacas), then vanished from North America entirely.

Elephants soon overran the entire world except Australia. For once the world was dominated by a huge animal with a brain sufficiently large to enable it to carry on.

The elephants. The text's admiration for these animals is evident. They are singled out as exceptional, the first large-bodied animals with brains adequate to their size. Modern research confirms that elephants possess remarkable cognitive abilities: they recognize themselves in mirrors, mourn their dead, use and even modify tools, and display behavior suggesting complex social understanding.

No animal the size of an elephant could have survived unless it had possessed a brain of large size and superior quality. In intelligence and adaptation the elephant is approached only by the horse and is surpassed only by man himself.

Intelligence as a survival strategy. The dinosaurs had favored size over brains. The mammals would favor brains over size, or at least would require that size be matched by intelligence. The elephants represent an evolutionary experiment in combining the two: large bodies supported by large brains. They were successful. Of the approximately two dozen proboscidean species that existed at the beginning of the Late Pleistocene, only three survive today, but they have spread across two continents and survived the ice ages that eliminated so many of their contemporaries.

With each expansion of brain size, with each new capacity for memory and abstraction and social learning, the universe was drawing closer to something. Not merely smarter animals, but something qualitatively new: creatures who could wonder where they came from.

Yet intelligence was no guarantee of survival:

Of the fifty species of elephants in existence at the opening of this period, only two have survived.

The text counted two species. Today we recognize three: the African bush elephant, the African forest elephant (confirmed as a distinct species in 2001), and the Asian elephant. All are endangered. The great proboscidean diversity of the Cenozoic has dwindled to a handful of survivors, and those survivors are threatened by human activity.

About this time a notable thing occurred in western North America: The early ancestors of the ancient lemurs first made their appearance. While this family cannot be regarded as true lemurs, their coming marked the establishment of the line from which the true lemurs subsequently sprang.

The primates. Our own lineage, beginning its long journey toward humanity. The earliest primates were small creatures, similar to modern mouse lemurs, adapted to life in the forest canopy. While the most primitive stem primates (plesiadapiforms) appear in the North American fossil record, current evidence suggests that true crown primates likely originated in Asia, spreading rapidly to other continents. From these humble beginnings would eventually emerge the monkeys, the apes, and ultimately ourselves.

The text traces the primate lineage through a specific sequence: lemur ancestors giving rise to true lemurs, then to mid-mammals, then to primates proper, then finally to humans. Modern paleontology tells a somewhat different story, placing human origins in Africa rather than in North America, deriving us from African apes rather than from a separate primate lineage. But the basic insight—that humans emerged from a long evolutionary sequence of increasingly intelligent primates —remains sound.

The text describes the first step:

> *Slightly to the west of India, on land now under water and among the offspring of Asiatic migrants of the older North American lemur types, the dawn mammals suddenly appeared. These small animals walked mostly on their hind legs, and they possessed large brains in proportion to their size and in comparison with the brains of other animals.*

The dawn mammals. Bipedal. Large-brained. These were the ancestors that would eventually give rise to humans. The text emphasizes the same qualities that modern paleoanthropology identifies as crucial to human evolution: upright posture and expanded brains.

> *In the seventieth generation of this order of life a new and higher group of animals suddenly differentiated. These new mid-mammals, almost twice the size and height of their ancestors and possessing proportionately increased brain power, had only well established themselves when the Primates, the third vital mutation, suddenly appeared.*

Sudden differentiation. The text describes evolution as proceeding through discrete jumps rather than gradual accumulation of small changes. This echoes the pattern described earlier in marine life, where new forms appeared suddenly. Whether one interprets this as evidence of divine intervention or as a description of what we now call punctuated equilibrium, the pattern is striking.

> *At this same time, a retrograde development within the mid-mammal stock gave origin to the simian ancestry; and from that day to this the human branch has gone forward by progressive evolution, while the simian tribes have remained stationary or have actually retrogressed.*

The separation of humans and apes. Modern genetics has clarified this relationship: humans and chimpanzees share a common ancestor that lived approximately six to seven million years ago. We did not descend from modern apes; we share ancestors with them. Both lineages have been evolving ever since the split, though the human lineage has undergone more dramatic changes in brain size and behavior.

<hr>

The age of mammals had produced elephants and horses, whales and bats, primates and the ancestors of humans. The Cenozoic era, from the extinction of the dinosaurs to the present day, had witnessed the transformation of a world dominated by reptiles into a world dominated by warm-blooded, nursing creatures whose young were born live and helpless, dependent on parental care for their survival.

And then the world grew cold:

The ice age is the last completed geologic period, the so-called Pleistocene, over two million years in length.

The warm greenhouse world of the early Cenozoic was giving way to something harsher: an era of glacial advances and retreats, of ice sheets that covered continents, of a climate that would test every species' ability to adapt. The mammals had proved their worth in the benign conditions of the Eocene and Miocene. Now they would face their greatest challenge.

The enforced migration of life before the advancing ice led to an extraordinary commingling of plants and of animals, and with the retreat of the final ice invasion, many arctic species of both plants and animals were left stranded high upon certain mountain peaks, whither they had journeyed to escape destruction by the glacier. And so, today, these dislocated plants and animals may be found high up

The ice shaped the world we know. The Great Lakes of North America are glacial scars. The fjords of Norway are drowned glacial valleys. The distribution of plants and animals across the Northern Hemisphere still reflects the migrations forced by advancing and retreating ice. And it was in this harsh, challenging environment that the human lineage would take its final steps toward becoming truly human.[1]

13

THE ICE AGES

*Y*our *early ancestors were born and bred in a stimulating, invigorating, and difficult environment.*

The ice came slowly at first.

High in the mountains of northern Canada, snow fell each winter and failed to melt completely each summer. Year by year, the residual snow accumulated. Decades passed. Centuries. Millennia. The snow compacted into ice, and the ice began to flow outward under its own weight, creeping down from the highlands like a frozen tide.

2,000,000 years ago the first North American glacier started its southern advance.

The beginning of the Pleistocene epoch (now dated to approximately 2.6 million years ago), the great ice age that would reshape the Northern Hemisphere and forge the human species in its frigid crucible. What followed was not a single frozen period but a complex rhythm of glacial advances and retreats, of ice sheets that grew and shrank and grew again, responding to subtle variations in Earth's orbit and axial tilt.

> *The northern regions of this world have experienced six separate and distinct ice invasions, although there were scores of advances and recessions associated with the activity of each individual ice sheet.*

Modern geology has revealed an even more complex picture. Deep-sea oxygen isotope records now identify over twenty distinct glacial-interglacial cycles during the Pleistocene, far exceeding the six major glaciations described here. The traditional models of four or six glaciations have been replaced by a more nuanced understanding of rhythmic climate oscillations operating on approximately 100,000-year periodicities. The Pleistocene was not uniformly cold. Between the major advances, interglacial periods brought warmth comparable to today's climate, sometimes even warmer. Forests returned to regions that had been buried under miles of ice. Animals migrated northward, following the retreating glaciers. Then the cold would return, and the cycle would begin again.

> *The ice in North America collected in two and, later, three centers. Greenland was covered, and Iceland was completely buried beneath the ice flow. In Europe the ice at various times covered the British Isles excepting the coast of southern England, and it overspread western Europe down to France.*

The geography of ice age Earth was dramatically different from today's. The British Isles were not islands but a peninsula of mainland Europe, connected by land that now lies beneath the North Sea. So much water was locked in glaciers that sea levels dropped by more than four hundred feet, exposing continental shelves around the world. The Bering land bridge connected Asia to North America, allowing animals and eventually humans to walk from one continent to the other.

The great ice sheets of this period were all located on elevated highlands, not in mountainous regions where they are found today. One half of the glacial ice was in North America, one fourth in Eurasia, and one fourth elsewhere, chiefly in Antarctica.

The Laurentide ice sheet that covered much of North America was comparable in size to the Antarctic ice sheet today—both containing roughly 26-27 million cubic kilometers of ice. North America's geography was ideally suited to ice sheet formation: broad highlands in the Canadian Shield, cold air masses sweeping down from the Arctic, moisture from the Great Lakes and Atlantic Ocean feeding constant snowfall.

In this invasion the three great ice sheets coalesced into one vast ice mass, and all of the western mountains participated in this glacial activity. This was the largest of all ice invasions in North America; the ice moved south over fifteen hundred miles from its pressure centers, and North America experienced its lowest temperatures.

Fifteen hundred miles of ice. Imagine standing at the southern edge of the glacier, somewhere in what is now Missouri or Illinois. Behind you, stretching to the horizon and beyond, a wall of ice rises hundreds of feet into the air. The cold radiating from it makes the air shimmer. Dust and debris embedded in the ice face give it a dirty gray appearance. Meltwater streams pour from its base, carrying sand and gravel that will eventually become the fertile soils of the Midwest.

The weight of this ice was almost incomprehensible. Up to two and a half miles thick in places, pressing down on the land with such force that the continental crust itself was depressed, pushed down into the underlying mantle. Even today, thousands of years after the ice melted, the land is still rising, rebounding from the burden it once carried. This process, called isostatic rebound, continues at rates of up to ten millimeters per year in some regions around Hudson Bay.

The date given here does not match modern geological evidence. The Last Glacial Maximum—when ice sheets reached their greatest extent—occurred approximately 26,000 to 19,000 years ago, with peak ice coverage around 21,000 years ago. The extent described, however, matches well with geological evidence. Terminal moraines—ridges of debris pushed up at the glacier's edge—mark the southern boundary of the ice across the American heartland. Long Island, Cape Cod, and the rolling hills of the upper Midwest are all glacial landforms, shaped by ice that melted thousands of years ago.

The Great Lakes. Superior, Michigan, Huron, Erie, Ontario: five inland seas holding approximately eighteen to twenty percent of the world's fresh surface water. All carved by glacial ice, all filling basins gouged from bedrock by the relentless grinding of mile-thick glaciers. The lakes did not exist before the ice ages. They are young features by geological standards, formed only as the last glaciers retreated.

The process of their formation was violent and dramatic. As the ice melted, vast quantities of water accumulated behind ice dams. When these dams broke, catastrophic floods swept across the landscape, carving channels, depositing debris, reshaping the terrain in days what normal erosion would take millennia to accomplish. The Channeled Scablands of Washington State, a bizarre landscape of dry waterfalls and scoured basalt, testify to one such megaflood, when glacial Lake Missoula burst its ice dam and released water at a rate roughly fifteen times greater than all the rivers in the world combined.

> *And Earth geologists have very accurately deduced the various stages of this development and have correctly surmised that these bodies of water did, at different times, empty first into the Mississippi valley, then eastward into the Hudson valley, and finally by a northern route into the St. Lawrence.*

This is a notable acknowledgment of scientific accuracy. The text affirms that human geologists have correctly reconstructed the drainage history of the Great Lakes, a complex story of shifting outlets as ice dams formed and broke, as isostatic rebound changed the topography, as the lakes drained first one way and then another before settling into their current configuration.

What did the ice ages mean for life?

> *The enforced migration of life before the advancing ice led to an extraordinary commingling of plants and of animals, and with the retreat of the final ice invasion, many arctic species of both plants and animals were left stranded high upon certain mountain peaks, whither they had journeyed to escape destruction by the glacier.*

Forced migration. As the ice advanced, entire ecosystems shifted southward. Boreal forests that had covered Canada retreated into what is now the American Midwest. Tundra vegetation spread across regions that had been temperate woodland. Animals moved with their habitats or perished.

When the ice retreated, the reverse migration occurred, but not perfectly. Some species, adapted to cold conditions, found themselves stranded on mountaintops as the lowlands warmed around them. These relict populations, isolated on their alpine refuges, persist to this day: arctic plants growing on the summits of the Appalachians, tundra insects surviving on peaks in New England, biological communities that testify to the dramatic climatic shifts of the recent geological past.

> *And so, today, these dislocated plants and animals may be found high up on the Alps of Europe and even on the Appalachian Mountains of North America.*

The ice age left its mark everywhere. Not just in the sculpted landscapes and carved lake basins, but in the distribution of species, in the patterns of genetic diversity, in the very composition of the biological communities that surround us. We live in a postglacial world, still recovering from the most recent advance, still shaped by forces that began over two million years ago.

But the ice did not fall equally on all lands:

> *Africa was little affected by the ice, but Australia was almost covered with the antarctic ice blanket.*

The glaciations affected different regions differently. Africa, mostly tropical, experienced changes in rainfall patterns rather than ice coverage. The Sahara expanded and contracted with the glacial cycles, sometimes even larger than today, sometimes dotted with lakes and supporting rich ecosystems.

The claim about Australia, however, requires significant correction: Australia remained largely ice-free during the Pleistocene, with glaciation limited to small alpine areas in Tasmania and the highest peaks of the Kosciuszko massif—less than one percent of the continent's surface. Geological mapping of glacial deposits confirms this limited extent. Australia's climate was affected by global cooling, but it was never close to being covered by ice.

And what of humans?

1,000,000 years ago Earth was registered as an inhabited world. A mutation within the stock of the progressing Primates suddenly produced two primitive human beings, the actual ancestors of mankind.

One million years ago. The appearance of humans, according to the text, in the midst of the ice ages. Modern paleoanthropology places the emergence of *Homo erectus* approximately 1.9 to 2 million years ago in Africa, with anatomically modern *Homo sapiens* appearing around 300,000 years ago. The text's date falls within the period of early human evolution, though the exact species and timing remain subjects of ongoing research.

The ice ages were not merely an obstacle that humans had to overcome; they were the forge in which humanity was shaped. The challenges of glacial climate—the need to adapt to changing conditions, the pressure to develop tools and strategies for survival—pushed human evolution in directions that might not have occurred in easier circumstances.

Modern anthropology generally supports this view, though with different emphasis. The fluctuating climates of the Pleistocene created a constantly changing environment that favored flexibility, intelligence, and cultural adaptation. Species that could learn, that could modify their behavior in response to changing conditions, that could develop new technologies to exploit new resources, had advantages over species locked into fixed behavioral patterns.

And the sole survivors of these Earth aborigines, the Eskimos, even now prefer to dwell in frigid northern climes.

The Eskimos, or as they prefer to be called, the Inuit. People who have mastered the art of living in the harshest cold environments on Earth, who have developed technologies and cultural practices perfectly adapted to Arctic conditions. The text suggests they represent a direct continuation of the cold-adapted populations that emerged during the ice ages.

Human beings were not present in the Western Hemisphere until near the close of the ice age. But during the interglacial epochs they passed westward around the Mediterranean and soon overran the continent of Europe.

The migration of humans across the globe, following the rhythms of the glacial cycles. During warm periods, humans spread northward into Europe and Asia. During cold periods, they retreated southward or adapted in place. The colonization of the Americas came late, when the Bering land bridge was exposed by lowered sea levels and humans could walk from Asia to Alaska. Current evidence suggests humans arrived in the Americas approximately 15,000 to 20,000 years ago, with some controversial sites suggesting even earlier dates.

In the caves of western Europe may be found human bones mingled with the remains of both tropic and arctic animals, testifying that man lived in these regions throughout the later epochs of the advancing and retreating glaciers.

The cave sites of Europe: Lascaux, Altamira, Chauvet. These are not merely archaeological sites but time capsules, preserving not only human bones and artifacts but the bones of the animals that humans hunted, painted, and feared. The same caves contain remains of woolly mammoths and reindeer alongside hippopotamus and rhinoceros—arctic species from glacial periods mingled with warmer-climate species from interglacials, documenting the dramatic climate oscillations of the Pleistocene.

The ice age is the last completed geologic period, the so-called Pleistocene, over two million years in length.

And then it ended:

> *35,000 years ago marks the termination of the great ice age excepting
> in the polar regions of the planet.*

The date given here is significantly earlier than modern estimates.
The Pleistocene epoch ended approximately 11,700 years ago, when
the Younger Dryas cold period gave way to the warmer Holocene.
The ice began its final retreat. The world began to warm. The great
ice sheets that had covered North America and Europe began to melt,
releasing floods of water, raising sea levels, drowning the land bridges
that had connected continents. A new era was beginning, the era in
which we still live: the Holocene, the age of human civilization.

> *This narrative, extending from the rise of mammalian life to the
> retreat of the ice and on down to historic times, covers a span of
> almost fifty million years. This is the last, the current, geologic period
> and is known to your researchers as the Cenozoic or recent-times era.*

The ice ages had lasted over two million years. They had shaped the
continents, carved the lakes, distributed the species, and driven the
migrations that would populate every corner of the globe. And in
their harsh embrace, they had forged a new kind of creature: one that
could think, plan, adapt, and ultimately transform the world. The long
prelude was over. The human story was about to begin in earnest.

The ice is gone now, mostly. But its legacy persists in us: in the adaptability that harsh climates demanded, in the problem-solving intelligence that survival required, in the social bonds that kept small bands
of primates alive through winters that would have killed any solitary
creature. We are children of the ice.[1]

14

THE THRESHOLD OF HUMANITY

he great event of this glacial period was the evolution of primitive man.

And so we arrive at the end of our journey through deep time, and at the beginning of another.

For nearly fourteen billion years the cosmos had been preparing. Galaxies had formed from primordial hydrogen. Stars had ignited, lived, and died, seeding space with the heavy elements necessary for rocky planets. Our sun had coalesced from the debris of a stellar encounter. Earth had formed from the material torn from the solar atmosphere, had cooled, and had developed oceans, an atmosphere, and the conditions necessary for life.

For over three billion years, life had been evolving. From the first simple cells in warm shallow seas, through the long age of marine invertebrates, through the conquest of land and the reign of reptiles and the age of mammals, evolution had been climbing toward something. Now, in the harsh crucible of the ice ages, it reached its culmination—primitive man.

The text traces human ancestry through a specific sequence: from early primate ancestors to dawn mammals to mid-mammals to primates to humans. Each transition is described as sudden, a mutation that produces a new type of creature distinctly different from its parents. The description of ancestors as "lemur types" reflects early twentieth-century terminology; modern primatology recognizes that while early primates superficially resembled some present-day prosimians, the human lineage runs through the haplorhine branch (tarsiers, monkeys, apes) rather than the strepsirrhine lineage that led to modern lemurs.

Bipedalism and encephalization. These two traits—walking upright and possessing large brains relative to body size—would prove to be the defining characteristics of the human lineage. Modern paleoanthropology confirms their importance, though the sequence and timing differ from the text's account. Bipedalism appears to have preceded significant brain expansion by several million years. The famous fossil Lucy, an *Australopithecus afarensis* from Ethiopia dated to 3.2 million years ago, walked upright but possessed a brain not much larger than a chimpanzee's.

themselves when the Primates, the third vital mutation, suddenly appeared.

Three sudden mutations. Three leaps in the evolutionary sequence. The text emphasizes discontinuity, the appearance of genuinely new forms rather than the gradual accumulation of small changes. This pattern echoes what we have seen throughout: evolution proceeding by jumps, new body plans appearing suddenly, missing links remaining forever missing because they never existed.

At this same time, a retrograde development within the mid-mammal stock gave origin to the simian ancestry; and from that day to this the human branch has gone forward by progressive evolution, while the simian tribes have remained stationary or have actually retrogressed.

The divergence of humans and apes. Modern genetics has clarified this relationship considerably since the text was written. We now know that humans and chimpanzees share approximately 98-99 percent of their DNA in directly comparable sequences (though when insertions, deletions, and structural variations are included, overall genomic similarity drops to roughly 95-96 percent). We diverged from a common ancestor approximately six to seven million years ago, and neither lineage has remained static since the split. But the basic insight stands: at some point, the ancestors of humans took a different path from the ancestors of modern apes, and that path led to something unprecedented.

A mutation within the stock of the progressing Primates suddenly produced two primitive human beings, the actual ancestors of mankind.

Two individuals. A pair. The text identifies specific first humans—not a population gradually becoming human, but two individuals who were human while their parents were not. This is perhaps the most distinctive claim in this account of human origins.

Modern evolutionary theory does not work this way. Species do not appear fully formed in single individuals or pairs. They emerge gradually from populations, as genetic changes accumulate over generations until the descendants differ enough from their ancestors to be considered a new species. There is no single first human in the scientific account, only a continuum of change with arbitrary lines drawn by later observers.

Yet the text insists on the reality of a first human moment. Not merely biological change but something more: the appearance of genuine personhood, of moral awareness, of the capacity for relationship with the divine.

> *This event occurred at about the time of the beginning of the third glacial advance; thus it may be seen that your early ancestors were born and bred in a stimulating, invigorating, and difficult environment.*

The ice ages as crucible. The text suggests purpose in the timing: humans did not appear in some gentle tropical paradise but in a world of glacial advances and harsh climate. The challenges of survival in such an environment would have favored intelligence, flexibility, social cooperation—all the traits that would eventually enable humans to dominate the planet.

Modern paleoanthropology increasingly supports the view that climatic variability played a crucial role in human evolution. The fluctuating climates of the Pleistocene created constantly changing environments that rewarded adaptability over specialization. Species locked into narrow ecological niches went extinct when conditions changed. Species capable of learning, of modifying their behavior, of developing new technologies, survived and flourished.

1,000,000 years ago Earth was registered as an inhabited world.

Registered. The language implies a cosmic bureaucracy, a system that tracks the status of worlds throughout the universe. Earth, having produced genuine will creatures, beings capable of moral choice, was now officially part of the community of inhabited planets.

The significance of this cannot be overstated. In the text's framework, we are not merely the accidental products of blind evolutionary processes. We are recognized participants in a cosmic order, creatures whose existence matters to beings far beyond our world.

About one million years ago the immediate ancestors of mankind made their appearance by three successive and sudden mutations stemming from early stock of the lemur type of placental mammal.

The text places human emergence at approximately one million years ago. Modern paleoanthropology dates the emergence of the genus *Homo* to approximately 2.8 million years ago, with *Homo erectus* appearing around 1.9 million years ago and anatomically modern *Homo sapiens* appearing approximately 300,000 years ago. The text's date of one million years falls within the long span of *Homo erectus* existence but does not correspond to a specific milestone recognized by current science. What matters more than the exact date is the recognition that humanity emerged from a long evolutionary process, that we are connected by descent to all other life on Earth, that our bodies and brains were shaped by the same forces of natural selection that shaped every other species. We are not separate from nature. We are its product, its culmination, and perhaps its first glimpse of self-awareness.

The first humans were brother and sister, who recognized their difference from their animal relatives and chose to separate themselves, to flee northward, to establish a new lineage distinct from the tribes of their birth.

What did that flight look like? Two figures moving through a landscape of sparse vegetation and harsh winds. They carried crude tools and each other's names—Sonta-an and Sonta-en, "loved by mother" and "loved by father." They knew only that they were different, that the others of their tribe were not like them, that the north might hold something better. They could not have known they were walking into history. They knew only that they had to go.

Two young creatures recognizing their own uniqueness, choosing to build something new, becoming the ancestors of all humanity. The story resonates. We are all related. We all share common ancestors. The differences that divide us are superficial compared to the kinship that unites us.

> *In many respects, they were the most remarkable pair of human beings that have ever lived on the face of the earth. This wonderful pair, the actual parents of all mankind, were in every way superior to many of their immediate descendants, and they were radically different from all of their ancestors, both immediate and remote.*

Superior to their descendants. This reverses the common assumption that evolution means constant improvement. The text suggests that the first humans possessed qualities that were subsequently diluted, that the purity of the original human type was compromised by interbreeding with less evolved populations.

> *The decision of the twins to flee from the Primates tribes implies a quality of mind far above the baser intelligence which characterized so many of their later descendants who stooped to mate with their cousins of the simian tribes.*

Decision. Choice. This is what sets humans apart: not merely larger brains or upright posture or opposable thumbs, but the capacity for genuine decision, for moral choice, for taking responsibility for one's own future. The first truly human act was a choice: to be different, to seek something better, to refuse to remain what circumstances had made them.

This capacity for choice, for self-determination, for transcending biological imperatives through conscious decision, remains the distinctive mark of humanity. Other animals are intelligent. Other animals use tools, communicate, form social bonds, even seem to mourn their dead. But humans alone, as far as we know, can contemplate their own existence, question their own purposes, and choose their own paths. Humans alone can decide who they want to become.

And so the story of "Before Humans" comes to an end, at the moment when humans begin.

We have traveled from the eternal center of all things to the birth of our galaxy, from the ignition of our sun to the formation of our planet, from the first stirrings of life in ancient seas to the emergence of creatures who could look up at the stars and wonder what they were.

The journey took billions of years. It required the cooperation of cosmic forces beyond human comprehension: gravity and electromagnetism, strong and weak nuclear forces, the mysterious processes by which energy becomes matter and matter organizes itself into ever more complex forms. It required patience—the patience of a universe willing to wait billions of years for conditions to be just right.

What a story it has been. Nebulae spinning into stars. Stars forging heavy elements in their nuclear furnaces. Planets coalescing from cosmic debris. Oceans forming on a cooling world. Life appearing in warm shallow seas and spreading, diversifying, complexifying, moving toward something it could not know.

Trilobites crawled the seafloors. Fish developed backbones. Amphibians hauled themselves onto land. Reptiles conquered the continents and grew to monstrous sizes before vanishing in a cosmic catastrophe. Mammals emerged from the shadows and inherited the Earth. And finally, in the harsh world of the ice ages, creatures appeared with the capacity to contemplate their own existence.

This narrative, extending from the rise of mammalian life to the retreat of the ice and on down to historic times, covers a span of almost fifty million years.

But the full story stretches back far further: billions of years to the formation of Earth, hundreds of billions of years to the birth of our nebula, and ultimately to Paradise itself, the eternal center from which all things derive.

And now here we are, creatures born of stardust, able to contemplate our origins. The cosmos looking back at its own beginnings, pondering the stars that made us, and the divine hand that made the stars.

None of this was accidental. Purpose runs through the entire story from beginning to end. The universe was designed to produce beings capable of knowing and loving and growing toward an infinite destiny. Science describes what happened and how. It does not ask why.

But perhaps that question—the question of meaning—is itself evidence that meaning exists. We are the creatures who ask why. We are the ones who seek purpose, who demand that existence make sense, who refuse to accept that all of this, the grandeur and the horror and the beauty and the pain, is simply atoms bouncing against atoms in patterns that signify nothing.

Every chapter of this book has contained moments where the text anticipated scientific discoveries that lay decades in the future, and moments where it contradicted what science would later establish.

Continental drift, described as fact before geologists accepted it. Quantum mechanics, acknowledged before its implications were fully understood. Timeline discrepancies that cannot be reconciled with radiometric dating. Details about dinosaurs and mammals and humans that modern research has refined or refuted.

What are we to make of a text that is sometimes prescient and sometimes mistaken, sometimes profound and sometimes puzzling? Perhaps we are to read it as its authors intended: not as a substitute for scientific investigation, but as a framework for understanding, a story that gives meaning to the facts, a narrative that places humanity within a cosmic context of purpose and significance.

We are human. We have crossed the threshold. Whatever comes next, the long story that led to this moment has given us the capacity to face it with courage, with creativity, and with hope.

The drama of world-making is complete. The drama of human history has just begun.[1]

We have traveled a long way together.

From the eternal stillness at the heart of all things to the spinning fire of newborn galaxies. From clouds of hydrogen collapsing into stars to a dark giant passing too close to our young sun. From a molten world wrapped in steam to oceans teeming with trilobites. From the first green things crawling onto barren shores to forests of giant ferns echoing with the footsteps of dinosaurs. From small furry creatures hiding in the shadows to beings who could look up at the stars and wonder what they were.

Fourteen billion years, compressed into fourteen chapters. The story of everything that happened before us, so that we could happen.

What are we to make of it?

The materialist says: nothing. It is atoms and void, cause and effect, a universe that doesn't know we exist and wouldn't care if it did. We are brief arrangements of matter that will soon dissolve back into the cosmic background. Enjoy it while it lasts. There is no meaning except what we invent.

There is a certain bleak courage in facing a meaningless universe and choosing to live anyway, to love anyway, to create meaning in the teeth of indifference.

But I do not believe it.

I have spent twenty years studying the account presented in this book, and I have come to believe that the universe is not indifferent. That the story has a Storyteller. That the drama of world-making was never aimless but always moved toward something—toward creatures who could know their Creator, who could choose goodness freely, who could participate consciously in a cosmic adventure barely begun.

No one can prove this. The deepest questions—Why is there something rather than nothing? Why is the universe comprehensible? Why do we hunger for meaning?—do not yield to proof. They yield to experience, to intuition, to the quiet voice that speaks when we are still enough to listen.

What I can say is that the story makes more sense if it has a point. The fine-tuning of physical constants that permits matter to exist. The precise conditions that allow stars to form heavy elements. The narrow habitable zone where water can remain liquid. The long chain of improbable events that led from single cells to consciousness. None of this proves purpose. But all of it is consistent with purpose. And purpose is what our hearts demand.

I wrote this book because I believe the story deserves to be told.

Not as dry science, though science has much to teach us. Not as religious doctrine, though faith has its place. But as narrative—as drama —as the grandest tale ever told, starring the cosmos itself and leading, improbably but inevitably, to you.

You are not a spectator to this story. You are its product and perhaps its purpose. Every atom in your body was forged in the heart of a dying star. Every cell reflects billions of years of evolution.

If the story has a Storyteller, you are not an accident. You are known. You are intended. You are loved by whatever force or being set this whole extraordinary chain of events in motion.

For now, it is enough to know where we came from. To feel the weight of billions of years pressing behind this present moment. To recognize that we are not accidents but inheritors of a cosmic legacy beyond comprehension.

The drama of world-making is complete.

What we make of it is up to us.

FROM THE AUTHOR

Thank you for reading. This book is the culmination of twenty years of seeking, study, and reflection.

If you're willing to share your thoughts, reader reviews make a meaningful difference for independent authors. Thank you so much.

APPENDIX

The narrative in this book is drawn primarily from *The Urantia Book*, a 2,097-page work first published in 1955 that claims to be a revelation presented by celestial beings to clarify and expand human understanding of cosmic reality and our place within it. This book exclusively cites the 1955 edition, which is in the public domain. The term "source text" is used in the following citations when referencing a paper from this book.

You may have never heard of it. Or you may have heard of it and dismissed it. That's fine.

Throughout this narrative, I have also drawn on findings from modern science—cosmology, geology, paleontology, biology—to provide context, to highlight where the text's claims align with subsequent discoveries, and to acknowledge honestly where they diverge. The endnotes for each chapter cite scientific sources alongside the revelatory material, because truth has nothing to fear from investigation.

What matters is whether the information resonates as true, whether it elevates your understanding, whether it helps you see the story of our world with fresh wonder. *The Urantia Book* has its critics and its devoted students. It's been called everything from the most important spiritual text of the modern era to elaborate fiction. I'm not asking you to accept it blindly. I'm asking you to read it and then decide if you think it is true.

I do. And I have read it countless times.

The source is less important than the truth it contains. And if you want to explore further, *The Urantia Book* is available online and in print.

About the Source Material

The Urantia Book is a comprehensive revelatory work covering a wide variety of subjects including cosmology, philosophy, history, and spirituality. It describes the nature of reality from the perspective of celestial beings and provides detailed information about the structure of the universe, the nature of God, the purpose of human existence, and the journey of the soul after death.

The book is organized into 196 papers grouped into four parts:

Part I: The Central and Superuniverses — The nature of God, the geography of the cosmos, the hierarchies of celestial beings

Part II: The Local Universe — Our region of space, its administration, the origins of life

Part III: The History of Urantia (Earth) — Planetary formation, the evolution of life, human history from earliest times through the twentieth century

Part IV: The Life and Teachings of Jesus — A detailed account of Jesus's entire life, from birth through resurrection

This book, *Before Humans*, draws primarily on Parts I, II, and III—the papers dealing with cosmic structure, planetary formation, and the long history of life on Earth.

A Note on Method

While I do, at times, exercise creative license in presenting this material as flowing narrative, my intention is never to stray from what the book discloses as revelatory fact. I have translated archaic phrasing into contemporary language, combined material from multiple papers into coherent chapters, and added scientific context that the original authors could not have included. But the story itself—the sequence of events, the key claims, the underlying vision—comes directly from the source.

Any mistakes are mine to own and correct.

The following references are organized by chapter to help readers locate the source material corresponding to specific content in this book.

NOTES

INTRODUCTION

1. Paper 101, Section 4: The Real Nature of Religion, The Limitations of Revelation

1. THE ARCHITECTURE OF INFINITY

1. The epigraph and embedded quotations in this chapter derive from Papers 11, 12, and 15 of the source text. The problems with Big Bang cosmology noted here are acknowledged in the scientific literature: the horizon problem, flatness problem, and magnetic monopole problem led to the development of inflationary theory (Guth, "Inflationary universe: A possible solution to the horizon and flatness problems," Physical Review D, 1981); the cosmological constant problem remains unresolved (Weinberg, "The Cosmological Constant Problem," Reviews of Modern Physics, 1989); the CMB alignment anomaly, later dubbed the "axis of evil," was first noted by de Oliveira-Costa et al. (2004) and Schwarz et al. (2004), with Land & Magueijo coining the term in Physical Review Letters (2005); and early reports from the James Webb Space Telescope identified unexpectedly massive galaxies at high redshift (Labbé et al., "A population of red candidate massive galaxies ~600 Myr after the Big Bang," Nature, 2023), though subsequent studies have explained some findings while confirming others remain anomalous. The precise dimensions of Paradise (ellipsoid, one-sixth longer north-south than east-west, flat with depth one-tenth of width) appear in Paper 11, Section 2. The space respiration cycle appears in Paper 11, Section 6. The charted cosmic path of our solar system appears in Paper 15, Section 1. The expansion of the universe was first described by Lemaître in 1927 and confirmed with additional data by Hubble in "A Relation between Distance and Radial Velocity among Extra-Galactic Nebulae," Proceedings of the National Academy of Sciences, 1929; the International Astronomical Union renamed the relationship the "Hubble-Lemaître Law" in 2018.

2. THE NATURE OF MATTER

1. The epigraph and quotations in this chapter derive from Paper 42 of the source text. The concept of ultimatons as subelectronic particles has no current experimental support; the Standard Model of particle physics treats electrons as fundamental leptons with no internal structure. However, the text's assertion that energy is absorbed and released in discrete quanta aligns precisely with quantum mechanics, formalized by Max Planck in a paper presented to the German Physical Society on December 14, 1900. Einstein extended Planck's work in 1905 by proposing that light itself consists of discrete quanta (photons), explaining the photoelectric effect. The claim regarding vacuum energy density resonates with modern quantum field theory, which describes the vacuum ground state as having non-zero energy with

measurable effects such as the Casimir force (experimentally confirmed by Lamoreaux, 1996). The ten grand divisions of matter appear in Paper 42, Section 3. For an accessible introduction to quantum mechanics and particle physics, see Brian Greene, The Elegant Universe (W.W. Norton, 1999), or the Particle Data Group reviews at pdg.lbl.gov.

3. BIRTH OF A NEBULA

1. The epigraph derives from Paper 15, Section 4 of the source text. The primary narrative draws on Paper 57, Sections 1-3. The timeline presented (nebular initiation approximately 875 billion years ago, first sun born approximately 500 billion years ago) differs dramatically from current cosmological consensus, which dates the observable universe at approximately 13.8 billion years (Planck Collaboration, "Planck 2018 results. VI. Cosmological parameters," Astronomy & Astrophysics, 2020). The description of nebulae as birthplaces of stars aligns with modern astrophysics; we observe stellar formation occurring in molecular clouds and nebulae throughout our galaxy. However, the mechanism described (centrifugal ejection of stars from spinning nebulae) differs from current stellar formation theory, which emphasizes gravitational collapse within nebulae (McKee & Ostriker, "Theory of Star Formation," Annual Review of Astronomy and Astrophysics, 2007). The Orion Nebula, Eagle Nebula, and Carina Nebula mentioned are well-documented stellar nurseries accessible to amateur and professional observation. The Orion Nebula (M42) is located approximately 1,344 light-years from Earth and is the closest massive star-forming region to our solar system (Ricci, Robberto, & Soderblom, "The Hubble Space Telescope/Advanced Camera for Surveys Atlas of Protoplanetary Disks in the Great Orion Nebula," The Astronomical Journal, 2008).

4. A STAR IS BORN

1. The epigraph derives from Paper 57, Section 3, describing the birth of suns from the Andronover nebula. The chapter draws primarily on Paper 57, Sections 4-5 of the source text. Modern solar science dates our sun at approximately 4.6 billion years old (Bouvier & Wadhwa, "The age of the Solar System redefined by the oldest Pb–Pb age of a meteoritic inclusion," Nature Geoscience, 2010), compared to the estimate of approximately 6 billion years, a discrepancy of roughly 30%, though impressively close for pre-radiometric estimates. The sun's classification as a G-type main-sequence star and its metallicity are well-established (Asplund et al., "The Chemical Composition of the Sun," Annual Review of Astronomy and Astrophysics, 2009). The sunspot cycle averages approximately 11 years, with individual cycles ranging from about 9 to 14 years (Hathaway, "The Solar Cycle," Living Reviews in Solar Physics, 2015). The encounter hypothesis for solar system formation was proposed by Chamberlin & Moulton (1905) and developed by Jeans (1917) and Jeffreys (1918), but was later superseded by the nebular hypothesis.

5. THE FORGING OF WORLDS

1. The epigraph derives from Paper 57, Section 6 of the source text. This chapter draws primarily on Paper 57, Sections 5-7. The description of tidal locking and the moon's future disruption within Earth's Roche limit aligns with modern celestial mechanics. The Roche limit concept was developed by Édouard Roche in 1848. Saturn's rings are now believed to be relatively young—perhaps 100 to 400 million years old—and likely the result of satellite disruption (Cassini mission data; Wisdom et al., Science, 2022). Jupiter radiates approximately 2.1 times more energy than it receives from the sun, confirming ongoing contraction and cooling (Li et al., Nature Communications, 2018). The reference to a destroyed fifth planet corresponds to the now-discredited "disrupted planet hypothesis" for the asteroid belt; modern consensus holds that the asteroids are primordial material prevented from accreting by Jupiter's gravity (Petit et al., Icarus, 2001), though recent models such as the Nice model and Grand Tack hypothesis allow for more complex early solar system dynamics including planetary migration and possible embryo ejection. Mercury's 3:2 spin-orbit resonance was discovered via radar observations by Pettengill and Dyce in 1965, correcting the earlier belief that Mercury was tidally locked to the Sun.

6. WHEN EARTH WAS FIRE

1. The epigraph derives from Paper 57, Section 8 of the source text. This chapter draws on Papers 57 (Sections 7-8) of the source text. Modern planetary science confirms that Earth formed through accretion approximately 4.54 billion years ago, with the age determined primarily through meteorite analysis (Patterson, "Age of meteorites and the earth," Geochimica et Cosmochimica Acta, 1956; refined by Bouvier & Wadhwa, Nature Geoscience, 2010). The text's endorsement of radiometric dating ("The radium clock is your most reliable timepiece") is notable for its period. Evidence for early liquid water comes from oxygen isotope analysis of ancient zircons (Wilde et al., "Evidence from detrital zircons for the existence of continental crust and oceans on the Earth 4.4 Gyr ago," Nature, 2001; Valley et al., "A cool early Earth," Geology, 2002). The Hadean eon designation was formally ratified by the International Commission on Stratigraphy in 2022. The USGS records approximately 55 earthquakes per day of magnitude 4 or greater, with about 15-16 major earthquakes (magnitude 7+) occurring annually. Continental crust (density ~2.7 g/cm³) floats higher than oceanic crust (density ~3.0 g/cm³) due to isostatic equilibrium; while oceanic crust is continuously recycled through subduction, continental cratons preserve the oldest rocks on Earth.

7. PREPARING THE CRADLE

1. The epigraph derives from Paper 58, Section 1 of the source text. This chapter draws extensively on Paper 58 (Sections 1-5) of the source text. The evolutionary significance of blood plasma's ionic similarity to seawater was developed by Quinton (1897) and Macallum ("The Paleochemistry of the Body Fluids and Tissues," Physiological Reviews, 1926). Ozone layer science was established by

Chapman (1930) and Dobson, who invented the ozone spectrophotometer (1924) and for whom the Dobson Unit is named; depletion effects were documented by Molina and Rowland (1974), earning the Nobel Prize in Chemistry (1995). Core temperature estimates derive from seismological constraints and high-pressure experiments (Anzellini et al., "Melting of Iron at Earth's Inner Core Boundary," Science, 2013), confirming temperatures of approximately 5,700-6,000°C at the inner core boundary. Continental drift was proposed by Wegener (1912) but not accepted until plate tectonics (Vine & Matthews, 1963; Wilson, 1965). The statement that Earth's core temperature slightly exceeds the sun's surface temperature aligns closely with current estimates (~5,700-6,000°C core vs. ~5,500°C solar photosphere). The euphotic zone, where sufficient light penetrates for photosynthesis, extends to approximately 200 meters (656 feet) in clear ocean water.

8. THE DAWN OF LIFE

1. The epigraph derives from Paper 58, Section 4 of the source text. This chapter draws extensively on Paper 58 (Sections 4-6) of the source text. The text's timeline for life's origin (550 million years ago) conflicts significantly with current scientific evidence placing the earliest life at approximately 3.5-4.1 billion years ago (Schopf, "Microfossils of the Early Archean Apex Chert: New Evidence of the Antiquity of Life," Science, 1993; Bell et al., "Potentially biogenic carbon preserved in a 4.1 billion-year-old zircon," PNAS, 2015). The concept of punctuated equilibrium was proposed by Eldredge and Gould ("Punctuated equilibria: an alternative to phyletic gradualism," in Models in Paleobiology, ed. T.J.M. Schopf, 1972). Continental drift was proposed by Wegener (1912) but not accepted until plate tectonics theory emerged in the 1960s. The text's description of continental positions and movements predates this acceptance. Osmoregulation in fish is described in standard physiology texts (e.g., Evans et al., "The Multifunctional Fish Gill: Dominant Site of Gas Exchange, Osmoregulation, Acid-Base Regulation, and Excretion of Nitrogenous Waste," Physiological Reviews, 2005). Slime molds are now classified as protists within the supergroup Amoebozoa, separate from plants, animals, and fungi.

9. THE AGE OF THE SEAS

1. The epigraph derives from Paper 59, Section 1 of the source text. This chapter draws extensively on Paper 59 (Sections 1-4) of the source text. The text's timeline corresponds roughly to the Paleozoic era (538.8-251.9 million years ago), with the Cambrian Explosion establishing most major animal body plans within the first 25 million years. Trilobite evolution and diversity are well documented (Fortey, Trilobite: Eyewitness to Evolution, 2000); their calcite lenses could contain over 15,000 individual facets in some species. Brachiopod diversity peaked at over 12,000 fossil species across more than 5,000 genera, with approximately 300-450 living species today (Sepkoski, "A Compendium of Fossil Marine Animal Genera," Bulletins of American Paleontology, 2002). The Niagara Formation (Silurian dolomite/limestone, 430-425 million years ago) forms the caprock over which Niagara Falls cascades. Modern gastropods comprise approximately 65,000-80,000 species, the largest class of mollusks (Bouchet et al., "Revised Classification, Nomenclator and

Typification of Gastropod and Monoplacophoran Families," Malacologia, 2017). The emergence of vertebrates in the Ordovician is supported by fossil evidence of jawless ostracoderms (Janvier, Early Vertebrates, 1996). Giant Ordovician nautiloids such as Endoceras giganteum reached lengths approaching 6 meters (19 feet). The final quote echoes Byron, Childe Harold's Pilgrimage, Canto IV, Stanza 54, 1818.

10. THE CONQUEST OF LAND

1. The epigraph derives from Paper 58, Section 1 of the source text. This chapter draws on Papers 58 and 59 (especially Sections 5-6) of the source text. The colonization of land by plants is dated to approximately 470 million years ago (Ordovician; Wellman et al., "Fragments of the earliest land plants," Nature, 2003). The Carboniferous period (359-299 million years ago) produced major coal deposits; elevated oxygen levels peaked at 25-35% during the Late Carboniferous and Early Permian (Berner, "Atmospheric oxygen over Phanerozoic time," PNAS, 1999). Giant insects of the Carboniferous include Meganeura (griffinfly, 65-75cm wingspan) and Arthropleura (millipede, up to 2.6m based on 2021 Northumberland discovery). The discovery of Tiktaalik roseae was a collaborative effort co-led by Neil Shubin (University of Chicago), Edward Daeschler (Academy of Natural Sciences), and Farish Jenkins Jr. (Harvard); see Daeschler, Shubin & Jenkins, "A Devonian tetrapod-like fish and the evolution of the tetrapod body plan," Nature, 2006. Shubin's account of the discovery appears in Your Inner Fish (Pantheon, 2008). The earliest tetrapods such as Ichthyostega (~1.5m) and Acanthostega (~60cm) were relatively modest in size; giant amphibians like Prionosuchus (up to 9m) appeared later in the Permian. The Permian-Triassic extinction (~252 million years ago) eliminated an estimated 81-96% of marine species (Erwin, "The Permo-Triassic extinction," Nature, 1994). The Paleozoic era spanned approximately 290 million years (538.8-251.9 Ma).

11. THE REIGN OF REPTILES

1. The epigraph derives from Paper 60, Section 1 of the source text. This chapter draws extensively on Paper 60 (Sections 1-4) of the source text. The amniotic egg evolved in the Carboniferous, enabling reptiles to colonize terrestrial habitats (Benton, Vertebrate Palaeontology, 4th ed., 2014). Dinosaur-bird ancestry was confirmed through cladistic analysis and feathered dinosaur discoveries in China (Xu et al., Nature, 1999-2014). Argentinosaurus size estimates from Bonaparte & Coria, Ameghiniana, 1993; Patagotitan weight revised to approximately 55-57 tons per Otero & Carballido (2020). The Chicxulub impact was proposed by Alvarez et al. ("Extraterrestrial cause for the Cretaceous-Tertiary extinction," Science, 1980) and the crater confirmed by Hildebrand et al. (Geology, 1991); Glen Penfield first detected the anomaly in 1978. Modern dating places the K-Pg extinction at 66 million years ago. The text's statement that dinosaurs are not found in Australia is inaccurate; Australia has produced numerous dinosaur fossils including Australotitan cooperensis (one of the world's ten largest dinosaurs), Australovenator wintonensis, Muttaburrasaurus langdoni, and Diamantinasaurus matildae. Archaeopteryx lithographica was discovered in the Solnhofen limestone quarries

of Bavaria in 1861, announced by Hermann von Meyer. Its significance as a transitional form is discussed in Wellnhofer, Archaeopteryx: The Icon of Evolution (Pfeil, 2009).

12. THE AGE OF MAMMALS

1. The epigraph derives from Paper 61, Section 3 of the source text. This chapter draws extensively on Paper 61 of the source text. Modern paleontology places mammalian origins in the Triassic (~225 million years ago), with mammals evolving from synapsids (specifically cynodonts), not from dinosaurs; synapsids and archosaurs (dinosaur ancestors) diverged approximately 315 million years ago (Rose, The Beginning of the Age of Mammals, 2006). Whale evolution from land mammals is documented by Gingerich et al. ("Origin of Whales from Early Artiodactyls," Science, 2001) and Thewissen et al. (Nature, 2007); the land-to-sea transition took approximately 10-11 million years, from Pakicetus (~50 Ma) to fully aquatic Basilosaurus (~40 Ma). Horse evolution in North America is detailed in MacFadden (Fossil Horses, 1992); Hyracotherium stood 30-60 cm tall with four toes on front feet and three on hind feet. Elephant cognitive abilities are reviewed in Plotnik et al. ("Self-recognition in an Asian elephant," PNAS, 2006); tool use documented by Hart et al. (2001) and mourning behavior by Douglas-Hamilton et al. (2006). Human-chimpanzee divergence is dated to ~6-7 million years ago based on molecular clock studies (Langergraber et al., PNAS, 2012). Proboscidean diversity at the beginning of the Late Pleistocene was approximately 23 species (not 50); three elephant species survive today. The Oligocene epoch began approximately 33.9 million years ago. While stem primates (plesiadapiforms) first appear in North American fossil record, true crown primates (euprimates) likely originated in Asia.

13. THE ICE AGES

1. The epigraph derives from Paper 61, Section 6 of the source text. This chapter draws extensively on Paper 61 (Sections 5-7) of the source text. Modern glaciology recognizes over twenty glacial-interglacial cycles during the Pleistocene, far exceeding the six described in the source text; the Lisiecki & Raymo benthic δ18O stack ("A Pliocene-Pleistocene stack of 57 globally distributed benthic δ18O records," Paleoceanography, 2005) provides the standard reference for Pleistocene climate oscillations. The Last Glacial Maximum is dated to approximately 26,000-19,000 years ago, not 150,000 years as stated in the text; the 150,000-year date roughly corresponds to the penultimate (Saalian/Illinoian) glaciation. The Pleistocene ended approximately 11,700 years ago, not 35,000 years ago. Great Lakes formation is documented in Larson & Schaetzl ("Origin and Evolution of the Great Lakes," Journal of Great Lakes Research, 2001). Isostatic rebound continues at rates of up to 10-13 mm/year in the Hudson Bay region (Sella et al., "Observation of glacial isostatic adjustment in 'stable' North America," Geophysical Research Letters, 2007). The Laurentide and Antarctic ice sheets were comparable in size at LGM, both containing approximately 26-27 million km³ of ice. Maximum ice thickness reached approximately 3-4 km (1.9-2.5 miles), not three miles. Australia was not "almost covered" by ice; glaciation was limited to approximately 7,000 km² in Tasmania and the Kosciuszko region, less than 0.1% of the continent. Human

arrival in the Americas is currently dated to approximately 15,000-20,000 years ago based on pre-Clovis archaeological evidence. Lake Missoula flood peak discharge (~17-20 million m³/s) exceeded total global river discharge (~1.2 million m³/s) by a factor of approximately 15.

14. THE THRESHOLD OF HUMANITY

1. The epigraph derives from Paper 61, Section 6 of the source text. This chapter draws on Papers 61, 62, and 63 of the source text. Modern paleoanthropology dates the emergence of the genus Homo to approximately 2.8 million years ago based on the LD 350-1 specimen from Ledi-Geraru, Ethiopia (Villmoare et al., "Early Homo at 2.8 Ma from Ledi-Geraru, Afar, Ethiopia," Science, 2015). Anatomically modern Homo sapiens appeared approximately 300,000-315,000 years ago based on fossils from Jebel Irhoud, Morocco (Hublin et al., "New fossils from Jebel Irhoud, Morocco and the pan-African origin of Homo sapiens," Nature, 2017). Human-chimpanzee genetic similarity is approximately 98-99% in directly alignable single-nucleotide sequences, though overall genomic similarity drops to ~95-96% when insertions, deletions, and structural variations are included (Chimpanzee Sequencing and Analysis Consortium, "Initial sequence of the chimpanzee genome and comparison with the human genome," Nature, 2005). The role of climate variability in human evolution is discussed in Potts ("Environmental hypotheses of hominin evolution," Yearbook of Physical Anthropology, 1998) and deMenocal ("Climate and Human Evolution," Science, 2011). The "variability selection hypothesis" proposes that environmental instability, rather than any single habitat type, drove key human adaptations. Bipedalism preceded significant brain expansion by approximately 4-5 million years: evidence for upright walking appears 6-7 million years ago (Orrorin, Sahelanthropus), while major brain expansion began only around 2 million years ago with early Homo. The description of human ancestors as "lemur types" reflects early terminology; the human lineage runs through the haplorhine branch of primates rather than the strepsirrhine lineage that led to modern lemurs.

ABOUT THE AUTHOR

Michael Vincent spent years searching for answers that religion couldn't provide. Then he discovered *The Urantia Book*—a dense revelation that answered his questions with a coherence he'd never encountered. His work translates this complex material into books anyone can absorb.

Before Humans is part of that project. Other books cover angels, history, Jesus, philosophy, and spirituality.

michaelvincentauthor.com

instagram.com/michaelvincent_author

facebook.com/HavonaPress

tiktok.com/@michael.vincent.author

youtube.com/@MichaelVincent-Author

amazon.com/author/havona-press